Advances in Gene Biotechnology

Advances in Gene Biotechnology

Shahzad Anwer

RANDOM
RANDOM PUBLICATIONS
NEW DELHI (INDIA)

Advances in Gene Biotechnology

ISBN 978-93-5111-552-6

Published in 2015 in India by

RANDOM PUBLICATIONS

4376-A/4B, Gali Murari Lal, Ansari Road
New Delhi-110 002
Phone : +9111-43580356, 011-23289044, 011-43142548
e-mail: sales@randompublications.com,
info@randompublications.com, randomexports@gmail.com

Reprinted 2026

Type Setting by : Friends Media, Delhi-110089

Preface

Gene Biotechnology presents tools that researchers and students need to understand and apply today's biotechnology techniques. Many of the currently available books in molecular biology contain only protocol recipes, failing to explain the principles and concepts behind the methods.

Gene technology sits within the broader area of biotechnology – the use of living things to make or change products. Humans have been using biotechnology for centuries in activities ranging from plant and animal breeding through to brewing and baking.

As our understanding of how living things function, grow and reproduce increases, modern biotechnology creates new opportunities for food and fibre production.

Modern biotechnology includes the discovery of genes (genomics), understanding how genes function and interact (functional genomics), discovery of natural DNA markers to select more efficient plants and animals and genetic modification or genetic engineering.

Biotechnology is a broad term that covers the practical use of biological systems to produce goods and services. It encompasses the transformation of materials by micro-organisms (eg. fermentation), methods of propagation, such as plant cloning or g r a fting, and may involve genetic alteration through methods such as selective breeding. Recent advances in biotechnology provide ways of introducing very precise changes to genetic material that allow, for the first time, the transfer of properties of a single gene from one organism to another.

These new techniques, commonly referred to as "gene technology", involve the modification of organisms by the direct incorporation (or deletion) of one or more genes to introduce or alter a specific characteristic or characteristics. Organisms created using gene technology techniques are commonly referred to as 'genetically modified organisms' (GMOs).

This book on gene biotechnology is being prepared by keeping in view the requirement of the students and researchers in biotechnology, genetic engineering and other allied fields.

I would like to thank my team for standing beside me throughout my career and writing this book. My special thanks go to "Random Publications" who have published the book.

– *Shahzad Anwer*

Contents

1

Gene Transfer Methods in Plants

To add a desired trait to a crop, a foreign gene (transgene) encoding the trait must be inserted into plant cells, along with a "cassette" of additional genetic material. The cassette includes a DNA sequence called a "promoter," which determines where and when the foreign gene is expressed in the host, and a "marker gene" that allows breeders to determine which plants contain the inserted gene by screening or selection. For example, marker genes may render plants resistant to antibiotics that are not used medically (*e.g.*, agromycin, canamycin) or tolerant to certain herbicides.

Two methods are used to transfer foreign genes into plants. The first method involves the use of a plant pathogen called Agrobacterium tumefaciens, which causes crown gall disease in many species. This bacterium has a plasmid, or loop of non-chromosomal DNA, that contains tumor-inducing genes (T-DNA), along with additional genes that help the T-DNA integrate into the host genome.

For genetic engineering purposes, Agrobacterium must first be "disarmed" so that it does not make the plant sick. This is done by removing most of the T-DNA while leaving the left and right border sequences, which integrate a foreign gene into the genome of cultured plant cells.

The second delivery method is a "gene gun," which fires gold particles carrying the foreign DNA into plant cells. Some of these particles pass through the plant cell wall and enter the cell nucleus, where the transgene integrates itself into the plant chromosome. Because both methods of gene transfer are fairly random, one must screen for the plant cells that contain the foreign gene.

GENES IN DEVELOPMENT

Gene regulation helps individual cells within an organism function in a specialized way. Other regulatory mechanisms coordinate the genes that determine how cells develop. All of the specialized cells in an organism, including those of the skin, muscle, bone, liver, and brain, derive from identical copies of a single fertilized egg cell. Each of these cells has the exact same DNA as the original cell, even though they have vastly different appearances and functions. Genes dictate how these cells specialize.

Early in an organism's embryonic development the overall body plan forms. Individual cells commit to a particular layer and region of the embryo, often migrating from one location to another to do so. As the organism grows, cells become part of a particular body organ or tissue, such as skin or muscle. Ultimately, most cells become highly specialized—not only to develop into a neuron rather than a muscle cell, for example, but to become a sensory neuron instead of a motor neuron. This process of specialization is called differentiation. At each stage of the differentiation process, specific genes known as developmental control genes actively turn on and switch off the genes that differentiate cells.

One class of developmental control genes, known as homeotic genes, directs the formation of particular body parts. Activating one set of homeotic genes instructs part of an embryo to develop into a leg, for example, while another set initiates the formation of the head. If a homeotic gene becomes altered or damaged, an organism's body development can be dramatically disrupted. A change in a single gene in some insects, for instance, can cause a leg to grow where an antenna belongs.

Homeotic genes work by regulating the activity of other genes. Homeotic genes code for the production of a regulatory protein that can bind to DNA and thus affect the transcription of one or more genes. This enables homeotic genes to initiate or halt the development and specialization of characteristics in an organism.

Nearly identical homeotic genes have been identified in varied organisms, such as insects, worms, mice, birds, and humans, where they serve similar embryonic development functions. Scientists theorize that homeotic genes first appeared in a single ancestor common to all these organisms. Sometime in evolutionary history, these organisms diverged from their common ancestor, but the homeotic genes continued to be passed down through generations virtually unchanged during the evolution of these new organisms.

GENOME RELATIONSHIPS AND WIDE-HYBRIDIZATION

Domestication of crops is likely to have captured only part of the allele diversity present in the wild progenitor gene pools, leaving many desirable traits absent from the crop. In some cases, such as barley 3 *Hordeum spontaneum* or maize 3 teosinte, crops can be crossed easily with wild relatives and there are few barriers to the incorporation of novel alleles since the genomes are essentially homologous.

In contrast, recently evolved polyploids such as hexaploid wheat will have had few opportunities to enlarge the gene pool. This is reflected by RFLP studies of wheat, barley and maize which show that wheat is less polymorphic.

The frequency with which allopolyploid species occur in nature shows that the processes of pollen germination, correct pollen tube growth, fertilization of the egg cell, and embryo development are often conserved in related species. This can be exploited to synthesize new combinations of genomes. Such interspecific or intergeneric hybrids are often referred to as wide-hybrids or wide-crosses. Wide-hybrids can be used to enlarge crop gene pools for any character but are most valuable where suitable genetic variation does not exist within a species, or where screening for a desired character would be difficult, time consuming or expensive. Disease resistance or abiotic stress tolerances are the characters most frequently targeted for transfer.

Barriers to Wide-hybridization

The ease of producing wide-hybrids, and the ease with which genetic material may be transferred by backcrossing to a crop parent, varies greatly from cross to cross. Successful gene transfer may be restricted by prefertilization, postfertilization or chromosome instability barriers. These are not necessarily directly related to the degree of genetic divergence between two species and the genetical control of hybridization barriers may be simple.

Prefertilization barriers usually involve failure of pollen germination or inhibition of pollen tube growth, which may be at the surface of the stigma, in the stigma, or in the transmitting tissue of the ovary wall. The best studied incompatibility reactions affecting cereal wide-hybrids are due to the *Kr* genes of wheat where dominant alleles inhibit the growth of pollen tubes of other species at the base of the style.

These genes are best studied in crosses with rye and *Hordeum bulbosum*. No method is known for overcoming the effect of these genes, so wide-hybrids involving bread or durum wheat usually use genotypes with recessive, non-inhibiting alleles. In grasses there are, of course, well-characterized genes for self-incompatibility (the *S* and *Z* loci) but it is not known whether there is homology between these and the *Kr* loci or whether a similar mechanism of inhibition operates.

Postfertilization barriers leading to seed abortion are common in widehybrids. In cereals, this is frequently due to defective development and subsequent collapse of the endosperm. Embryos, however, are often viable, and hybrid plants can be recovered if the embryos are excised from developing seeds and cultured on suitable nutrient agar. In the *Triticeae*, using these techniques, a huge catalogue of wide-hybrids have been produced.

For example, hexaploid, tetraploid and diploid wheats and their wild relatives can be intercrossed but embryo rescue is usually unavoidable, especially if different ploidy levels are crossed. Similarly, ovule or embryo culture has been used to recover hybrid plants from *Brassica* crosses. In some crosses, for example between bread and durum wheats, hybrid necrosis kills the developing seedling. This is due to the epistatic interaction of independent genes in the respective species. Crosses within the separate gene pools are unaffected, probably because selection has eliminated necrotic combinations of alleles. Incompatibilities between nuclear genomes and the cytoplasm donated by the female parent may also contribute to lethality or sterility in certain crosses.

In vitro fertilization using isolated gametes, and the subsequent recovery of plants, is a means by which prefertilization and some postfertilization barriers might be overcome. Currently, *in vitro* fertilization has only been used successfully in maize, but such systems have potential for wide-crosses.

A further barrier to the production of wide-hybrids is the phenomenon of chromosome elimination. In such cases fertilization results in the formation of a hybrid zygote, but the complete chromosome complement of one of the parents is lost during subsequent mitotic divisions, leading to the production of a haploid embryo. The best characterized system is the cross between barley and its wild relative *Hordeum bulbosum*. However, the phenomenon occurs in a wide range of other cross combinations in the *Triticeae* including wheat 3 *H. bulbosum*, wheat 3 barley and wheat 3 maize, sorghum or pearl millet.

The mechanism of chromosome elimination is not understood but involves the inability of the chromosomes of one genome to successfully attach to the mitotic spindle. In some crosses, the occurrence of chromosome elimination is affected by the parental genomes since some barley 3 *H. bulbosum* crosses produce a high frequency of barley haploids while others give a high frequency of relatively stable hybrids. The genetical control of this regulation is probably fairly simple. Although chromosome elimination is detrimental to gene transfer, efficient haploid production methods benefit breeding and genetic analysis programmes and are extensively used by workers in these areas.

Utilization of Hybrid Plants

Once hybrid plants have been produced there are usually problems of sterility due to the inability of the hybrid to form viable gametes, particularly male gametes. Usually, this is due to a lack of chromosome pairing and chiasma formation at meiosis between the respective genomes and the subsequent formation of unbalanced gametes from random combinations of univalents.

This can occur even in cases where close homoeology between the genomes can be shown by comparative chromosome and genetic analysis. Clearly the mechanisms controlling the pairing of chromosomes can be so precise as to prevent homoeologous pairing between different genomes. Such mechanisms may be genetically simple, as in wheat where pairing between the A, B and D genomes of bread wheat is restricted by a few pairing control genes of which the most powerful is *Ph1* on chromosome 5B.

Fertility can be partially or fully restored in many hybrids by doubling the chromosome number with colchicine or an equivalent spindle inhibitor as this provides identical pairing partners for each chromosome. In such cases new allopolyploid 'species' can be established.

The major motivation for creating wide-hybrids is the introgression of useful genes into cultivated species. Generally, there are three levels at which the transfer of genetic material can be considered. Firstly, a whole genome can be introduced into an existing crop species to create a new allopolyploid species.

For example, the introduction of a diploid rye genome into durum or bread wheat to create hexaploid or octaploid triticales, respectively, has been highly successful. New genome combinations such as *Tritordeum* (the amphiploid between wheat 3 *Hordeum chilense*) are also being investigated as crops for specific environments.

Secondly, whole chromosomes can be introduced to create substitution or addition lines. Thirdly, translocations can be induced between homoeologous or non-homoeologous chromosomes so that only individual arms or small segments are introduced. The starting point for introgression is to backcross a wide-hybrid to the cultivated species.

In many cases this is possible using the hybrid as the female parent, even if the hybrid is highly sterile, since a proportion of egg cells contain a non-reduced or otherwise viable chromosome complement.

If there is homoeologous pairing between genomes, translocation products can be identified using *in situ* hybridization or molecular markers and stabilized by further backcrossing. Backcrosses can also be tested for the presence of the desired trait or traits in the 'alien' chromosome segment.

Combinations of cytological and marker techniques can then be used to characterize suitable lines. In some hybrids pairing between homoeologous genomes does not occur, or occurs too infrequently to be useful. In wheat, this can be overcome by using the *ph1* mutant or plants lacking chromosome 5B (carrying the *Ph1* pairing control gene). This allows higher levels of homoeologous chromosome pairing. Alternatively radiation treatments can be used to induce translocations. In such circumstances it is usually more effective to develop single chromosome addition or substitution lines prior to selecting recombinants, particularly as such stocks can confirm that the desired trait is carried by a single chromosome.

Plant breeders are generally interested in maintaining the characteristics of the cultivated species while adding one or more useful genes from the alien source. Generally the aim is to reduce the size of the alien chromosome segment as much as possible since donor species may have many alleles that are undesirable in the crop.

This is greatly assisted by the use of molecular markers which define the alien segment. *In situ* hybridization using total genomic DNA or species-specific repeats is also valuable because it reveals the number and size of the alien segments. *In situ* methods are also increasingly able to locate low-copy sequences on plant chromosomes, thereby linking the molecular and physical mapping.

Comparative mapping information is important for the manipulation of alien chromosome segments because of the importance of producing plants with genetically balanced genomes.

For example, substitution of rye chromosome 1 for a wheat group 1 chromosome is successful because these chromosomes show conserved synteny throughout their length.

In contrast, rye chromosome 7 is translocated with respect to the wheat chromosomes of groups 2, 4 and 5 and substitution of rye chromosome 7 for a wheat group 7 chromosome produces a plant with reduced doses of group 7 segments and extra doses of group 2, group 4 and group 5 segments.

Comparative mapping therefore provides a framework for the assembly of new chromosome segments in genetically balanced combinations. This principle also applies to other crop groups such as *Brassica*.

VECTOR-MEDIATED OR INDIRECT GENE TRANSFER

Among the various vectors used in plant transformation, the Ti plasmid of *Agrobacterium tumefaciens* has been widely used. This bacteria is known as "natural genetic engineer" of plants because these bacteria have natural ability to transfer T-DNA of their plasmids into plant genome upon infection of cells at the wound site and cause an unorganized growth of a cell mass known as crown gall.

Ti plasmids are used as gene vectors for delivering useful foreign genes into target plant cells and tissues. The foreign gene is cloned in the T-DNA region of Ti-plasmid in place of unwanted sequences. To transform plants, leaf discs (in case of dicots) or embryogenic callus (in case of monocots) are collected and infected with *Agrobacterium* carrying recombinant disarmed Ti-plasmid vector.

The infected tissue is then cultured (co-cultivation) on shoot regeneration medium for 2-3 days during which time the transfer of T-DNA along with foreign genes takes place. After this, the transformed tissues (leaf discs/calli) are transferred onto selection cum plant regeneration medium supplemented with usually lethal concentration of an antibiotic to selectively eliminate non-transformed tissues. After 3-5 weeks, the regenerated shoots (from leaf discs) are transferred to root-inducing medium, and after another 3-4 weeks, complete plants are transferred to soil following the hardening (acclimatization) of regenerated plants. The molecular techniques like PCR and southern hybridization are used to detect the presence of foreign genes in the transgenic plants.

Vectorless or Direct Gene Transfer: In the direct gene transfer methods, the foreign gene of interest is delivered into the host plant cell without the help of a vector. The methods used for direct gene transfer in plants are:

- *Chemical Mediated Gene Transfer:* Chemicals like polyethylene glycol (PEG) and dextran sulphate induce DNA uptake into plant protoplasts.Calcium phosphate is also used to transfer DNA into cultured cells.
- *Microinjection:* The DNA is directly injected into plant protoplasts or cells (specifically into the nucleus or cytoplasm) using fine tipped (0.5 - 1.0 micrometerdiameter) glass needle or micropipette. This method of gene transfer is used to introduce DNA into large cells such as oocytes, eggs, and the cells of early embryo.
- *Electroporation:* It involves a pulse of high voltage applied to

protoplasts/cells/ tissues to make transient (temporary) pores in the plasma membrane which facilitates the uptake of foreign DNA. The cells are placed in a solution containing DNA and subjected to electrical shocks to cause holes in the membranes. The foreign DNA fragments enter through the holes into the cytoplasm and then to nucleus.

- *Particle gun/Particle Bombardment:* In this method, the foreign DNA containing the genes to be transferred is coated onto the surface of minute gold or tungsten particles (1-3 micrometers) and bombarded onto the target tissue or cells using a particle gun (also called as gene gun/shot gun/microprojectile gun).The microprojectile bombardment method was initially named as biolistics by its inventor Sanford. Two types of plant tissue are commonly used for particle bombardment- Primary explants and the proliferating embryonic tissues.
- *Transformation:* This method is used for introducing foreign DNA into bacterial cells *e.g.* E. Coli. The transformation frequency (the fraction of cell population that can be transferred) is very good in this method. *E.g.* the uptake of plasmid DNA by E. coli is carried out in ice cold $CaCl_2$ (0-5°C) followed by heat shock treatment at 37-45°C for about 90 sec. The transformation efficiency refers to the number of transformants per microgram of added DNA. The $CaCl_2$ breaks the cell wall at certain regions and binds the DNA to the cell surface.
- *Conjuction:* It is a natural microbial recombination process and is used as a method for gene transfer. In conjuction, two live bacteria come together and the single stranded DNA is transferred via cytoplasmic bridges from the donor bacteria to the recipient bacteria.
- *Liposome Mediated Gene Transfer or Lipofection:* Liposomes are circular lipid molecules with an aqueous interior that can carry nucleic acids. Liposomes encapsulate the DNA fragments and then adher to the cell membranes and fuse with them to transfer DNA fragments. Thus, the DNA enters the cell and then to the nucleus. Lipofection is a very efficient technique used to transfer genes in bacterial, animal and plant cells.
- *Selection of Transformed Cells from Untransformed Cells:* The selection of transformed plant cells from untransformed cells is an important step in the plant genetic engineering. For this, a marker gene (*e.g.* for antibiotic resistance) is introduced into the plant along with the transgene followed by the selection of an appropriate selection medium (containing the antibiotic). The segregation and stability of the transgene integration and expression in the subsequent generations can be studied by genetic and molecular analyses (Northern, Southern, Western blot, PCR).

- *Transgenic Plants with Beneficial Traits:* During the last decades, a tremendous progress has been made in the development of transgenic plants using the various techniques of genetic engineering. The plants, in which a functional foreign gene has been incorporated by any biotechnological methods that generally are not present in the plant, are called transgenic plants. As per estimates recorded in 2002, transgenic crops are cultivated world-wide on about 148 million acres (587 million hectares) land by about 5.5 million farmers. Transgenic plants have many beneficial traits like insect resistance, herbicide tolerance, delayed fruit ripening, improved oil quality, weed control etc.

Some of the commercially grown transgenic plants in developed countries are: "Roundup Ready" soybean, 'Freedom II squash', 'High- lauric' rapeseed (canola), 'Flavr Savr' and 'Endless Summer' tomatoes. During 1995, full registration was granted to genetically engineered Bt gene containing insect resistant 'New Leaf' (potato), 'Maximizer' (corn), 'BollGard' (cotton) in USA. Some of the traits introduced in these transgenic plants are as follows:

- *Stress Tolerance:* Biotechnology strategies are being developed to overcome problems caused due to biotic stresses (viral, bacterial infections, pests and weeds) and abiotic stresses (physical actors such as temperature, humidity, salinity etc).
- *Abiotic Stress Tolerance:* The plants show their abiotic stress response reactions by the production of stress related osmolytes like sugars (*e.g.* trehalose and fructans), sugar alcohols (*e.g.* mannitol), amino acids (*e.g.* proline, glycine, betaine) and certain proteins (*e.g.* antifreeze proteins). Transgenic plants have been produced which over express the genes for one or more of the above mentioned compounds. Such plants show increased tolerance to environmental stresses. Resistance to abiotic stresses includes stress induced by herbicides, temperature (heat, chilling, freezing), drought, salinity, ozone and intense light. These environmental stresses result in the destruction, deterioration of crop plants which leads to low crop productivity. Several strategies have been used and developed to build ressitance in the plants against these stresses.
- *Herbicide Tolerance:* Weeds are unwanted plants which decrease the crop yields and by competing with crop plants for light, water and nutrients. Several biotechnological strategies for weed control are being used *e.g.* the over-production of herbicide target enzyme (usually in the chloroplast) in the plant which makes the plant insensitive to the herbicide. This is done by the introduction of a modified gene that encodes for a resistant form of the enzyme targeted by the herbicide in weeds and crop plants. Roundup Ready crop plants tolerant to herbicide-Roundup, is already being used commercially.

The biological manipulations using genetic engineering to develop herbicide resistant plants are:

- Over-expression of the target protein by integrating multiple copies of the gene or by using a strong promoter.,
- Enhancing the plant detoxification system which helps in reducing the effect of herbicide.,
- Detoxifying the herbicide by using a foreign gene.,
- Modification of the target protein by mutation.

Some of the examples are:

- *Glyphosate resistance*: Glyphosate is a glycine derivative and is a herbicide which is found to be effective against the 76 of the world's worst 78 weeds. It kills the plant by being the competitive inhibitor of the enzyme 5-enoyl-pyruvylshikimate 3- phosphate synthase (EPSPS) in the shikimic acid pathway. Due to it's structural similarity with the substrate phosphoenol pyruvate, glyphosate binds more tightly with EPSPS and thus blocks the shikimic acid pathway.

Certain strategies were used to provide glyphosate resistance to plants.

- It was found that EPSPS gene was overexpressed in Petunia due to gene amplification. EPSPS gene was isolated from Petunia and introduced in to the other plants. These plants could tolerate glyphosate at a dose of 2- 4 times higher than that required to kill wild type plants.
- By using mutant EPSPS genes- A single base substitution from C to T resulted in the change of an amino acid from proline to serine in EPSPS. The modified enzyme cannot bind to glyphosate and thus provides resistance.
- The detoxification of glyphosate by introducing the gene (isolated from soil organism- Ochrobactrum anthropi) encoding for glyphosate oxidase into crop plants. The enzyme glyphosate oxidase converts glyphosate to glyoxylate and aminomethylphosponic acid. The transgenic plants exhibited very good glyphosate ressitance in the field.

Phosphinothricin Resistance: Phosphinothricin is a broad spectrum herbicide and is effective against broad-leafed weeds. It acts as a competitive inhibitor of the enzyme glutamine synthase which results in the inhibition of the enzyme glutamine synthase and accumulation of ammonia and finally the death of the plant. The disturbace in the glutamine synthesis also inhibits the photosynthetic activity.

The enzyme phosphinothricin acetyl transferase (which was first observed in Streptomyces sp in natural detoxifying mechanism against phosphinothricin) acetylates phosphinothricin, and thus inactivates the herbicide. The gene encoding for phosphinothricin acetyl transferase (bar gene) was introduced in transgenic maize and oil seed rape to provide resistance against phosphinothricin.

ABIOTIC STRESSES IN TEMPERATURE, DROUGHT, AND SALINITY

The abiotic stresses due to temperature, drought, and salinity are collectively also known as water deficit stresses. The plants produce osmolytes or osmoprotectants to overcome the osmotic stress. The attempts are on to use genetic engineering strategies to increase the production of osmoprotectants in the plants.

The biosynthetic pathways for the production of many osmoprotectants have been established and genes coding the key enzymes have been isolated. *E.g.* Glycine betaine is a cellular osmolyte which is produced by the participation of a number of key enzymes like choline dehydrogenase, choline monooxygenase etc.

The choline oxidase gene from Arthrobacter sp. was used to produce transgenic rice with high levels of glycine betaine giving tolerance against water deficit stress. Scientists also developed cold-tolerant genes (around 20) in Arabidopsis when this plant was gradually exposed to slowly declining temperature. By introducing the coordinating gene (it encodes a protein which acts as transcription factor for regulating the expression of cold tolerant genes), expression of cold tolerant genes was triggered giving protection to the plants against the cold temperatures.

VARIETY OF INSECTS, MITES AND NEMATODES

A variety of insects, mites and nematodes significantly reduce the yield and quality of the crop plants. The conventional method is to use synthetic pesticides, which also have severe effects on human health and environment.

The transgenic technology uses an innovative and eco-friendly method to improve pest control management.About 40 genes obtained from microorganisms of higher plants and animals have been used to provide insect resistance in crop plants The first genes available for genetic engineering of crop plants for pest resistance were Cry genes (popularly known as Bt genes) from a *bacterium Bacillus thuringiensis.*

These are specific to particular group of insect pests, and are not harmful to other useful insects like butter flies and silk worms. Transgenic crops with Bt genes (*e.g.* cotton, rice, maize, potato, tomato, brinjal, cauliflower, cabbage, etc.) have been developed. This has proved to be an effective way of controlling the insect pests and has reduced the pesticide use.

The most notable example is Bt cotton (which contains CrylAc gene) that is resistant to a notorious insect pest Bollworm *(Helicoperpa armigera).* There are certain other insect resistant genes from other microorganisms which have been used for this purpose. Isopentenyl transferase gene from Agrobacterium tumefaciens has been introduced into tobacco and tomato.

The transenic plants with this transgene were found to reduce the leaf consumption by tobacco hornworm and decrease the survival of peach potato aphid. Certain genes from higher plants were also found to result in the synthesis of products possessing insecticidal activity. One of the examples is the Cowpea trypsin inhibitor gene (CpTi) which was introduced into tobacco, potato, and oilseed rape for develping transgenic plants. Earlier it was observed that the wild species of cowpea plants growing in Africa were resistant to attack by a wide range of insects. It was observed that the insecticidal protein was a trypsin inhibitor that was capable of destroying insects belonging to the orders Lepidoptera, Orthaptera etc. Cowpea trypsin inhibitor (CpTi) has no effect on mammalian trypsin, hence it is non-toxic to mammals.

SEVERAL STRATEGIES FOR ENGINEERING PLANTS FOR VIRAL RESISTANCE

There are several strategies for engineering plants for viral resistance, and these utilizes the genes from virus itself (*e.g.* the viral coat protein gene). The virus-derived resistance has given promising results in a number of crop plants such as tobacco, tomato, potato, alfalfa, and papaya. The induction of virus resistance is done by employing virus-encoded genes-virus coat proteins, movement proteins, transmission proteins, satellite RNa, antisense RNAs, and ribozymes.

The virus coat protein-mediated approach is the most successful one to provide virus resistance to plants. It was in 1986, transgenic tobacco plants expressing tobacco mosaic virus (TMV) coat protein gene were first developed. These plants exhibited high levels of resistance to TMV. The transgenic plant providing coat protein-mediated resistance to virus are rice, potato, peanut, sugar beet, alfalfa etc. The viruses that have been used include alfalfa mosaic virus (AIMV), cucumber mosaic virus (CMV), potato virus X (PVX), potato virus Y (PVY) etc.

FUNGAL AND BACTERIAL INFECTIONS

As a defence strategy against the invading pathogens (fungi and bacteria) the plants accumulate low molecular weight proteins which are collectively known as pathogenesis-related (PR) proteins. Several transgenic crop plants with increased resistance to fungal pathogens are being raised with genes coding for the different compounds.

One of the examples is the Glucanase enzyme that degrades the cell wall of many fungi. The most widely used glucanase is beta-1,4-glucanase. The gene encoding for beta-1,4 glucanase has been isolated from barley, introduced, and expressed in transgenic tobacco plants. This gene provided good protection against soil-borne fungal pathogen Rhizoctonia solani. Lysozyme degrades chitin and peptidoglycan of cell wall, and in this way fungal infection can be reduced. Transgenic potato plants with lysozyme gene providing resistance to Eswinia carotovora have been developed.

Delayed Fruit Ripening: The gas hormone, ethylene regulates the ripening of fruits, therefore, ripening can be slowed down by blocking or reducing ethylene production. This can be achieved by introducing ethylene forming gene(s) in a way that will suppress its own expression in the crop plant. Such fruits ripen very slowly (however, they can be ripen by ethylene application) and this helps in exporting the fruits to longer distances without spoilage due to longer-shelf life.

The most common example is the 'Flavr Savr' transgenic tomatoes, which were commercialized in U.S.A in 1994. The main strategy used was the antisense RNA approach. In the normal tomato plant, the PG gene (for the enzyme polygalacturonase) encodes a normal mRNA that produces the enzyme polygalacturonase which is involved in the fruit ripening.

The complimentary DNA of PG encodes for antisense mRNA, which is complimentary to normal (sense) mRNA. The hybridization between the sense and antisnse mRNAs renders the sense mRNA ineffective. Consequently, polygalacturonase is not produced causing delay in the fruit ripening. Similarly strategies have been developed to block the ethylene biosynthesis thereby reducing the fruit ripening. *E.g.* transgenic plants with antisense gene of ACC oxidase (an enzyme involved in the biosynthetic process of ethylene) have been developed. In these plants, production of ethylene was reduced by about 97 per cent with a significant delay in the fruit ripening. The bacterial gene encoding ACC deaminase (an enzyme that acts on ACC and removes amino group) has been transferred and expressed in tomato plants which showed 90 per cent inhibition in the ethylene biosynthesis.

INHERIT MALE STERILITY IN PLANT

The plants may inherit male sterility either from the nucleus or cytoplasm. It is possible to introduce male sterility through genetic manipulations while the female plants maintain fertility.

In tobacco plants, these are created by introducing a gene coding for an enzyme (barnase, which is a RNA hydrolyzing enzyme) that inhibits pollen formation. This gene is expressed specifically in the tapetal cells of anther using tapetal specific promoter TA29 to restrict its activity only to the cells involved in pollen production.

The restoration of male fertility is done by introducing another gene barstar that suppresses the activity of barnase at the onset of the breeding season. By using this approach, transgenic plants of tobacco, cauliflower, cotton, tomato, corn, lettuce etc. with male sterility have been developed.

Transgenic Plants as Bioreactors: Plants can be used as cheap chemical factories that require only water, minerals, sun light and carbon dioxide to produce thousands of sophisticated chemical molecules with different structures. By transferring the right genes, plants can serve as bioreactors to modified or new compounds such as amino acids, proteins, vitamins, plastics, pharmaceuticals (peptides and proteins), drugs, enzymes for food industry and so on. The transgenic plants as bioreactors have some advantages such as the cost of production is low, there is an unlimited supply, safe and environmental friendly and there is no scare of spread of animal borne diseases.

Tobacco is the most preferred plant as a transgenic bioreactor because it can be easily transformed and engineered. Tobacco is an excellent biomass producer with about 40 tons of fresh leaf production as against *e.g.* rice with 4 tons. The seed production is very high (approx. one million seeds per plant) and it can be harvested several times in a year. Some of the uses of transgenic plants are:

Improvement of Nutrient Quality: Transgenic crops with improved nutritional quality have already been produced by introducing genes involved in the metabolism of vitamins, minerals and amino acids. A transgenic Arabidopsis thaliana that can produce ten-fold higher vitamin E (alpha-tocopherol) than the native plant has been developed.

The biochemical machinery to produce a compound close in structure to alpha-tocopherol is present in A. thaliana. A gene that can finally produce alpha-tocopherol is also present, but is not expressed. This dormant gene was activated by inserting a regulatory gene from a bacterium which resulted in an efficient production of vitamin E. Glycinin is a lysine-rich protein of soybean and the gene encoding glycinin has been introduced into rice and successfully expressed.

The transgenic rice plants produced glycinin with high contents of lysine. Using genetic engineering Prof Potrykus and Dr. Peter Beyer have developed rice which is enriched in pro-vitamin A by introducing three genes involved in the biosynthetic pathway for carotenoid, the precursor for vitamin A.

The aim was to help millions of people who suffer from night blindness due to Vitamin A deficiency, especially whose staple diet is rice. The presence of beta-carotene in the rice gives a characteristic yellow/orange colour, hence this pro-vitamin A enriched rice is named as Golden Rice. The genetic engineering is also being used to improve the taste of food *e.g.* a protein 'monellin' isolated from an African plant (Dioscorephyllum cumminsii) is about 100,000 sweeter than sucrose on molar basis. Monellin gene has been introduced into tomato and lettuce plants to improve their taste.

Improvement of Seed Protein Quality: The nutritional quality of cereals and legumes has been improved by using biotechnological methods. Two genetic engineering approaches have been used to improve the seed protein quality. In the first case, a transgene (*e.g.* gene for protein containing sulphur rich amino acids) was introduced into pea plant (which is deficient in methionine and cysteine, but rich in lysine) under the control of seed-specific promoter. In the second approach, the endogenous genes are modified so as to increase the essential amino acids like lysine in the seed proteins of cereals.

These transgenic routes have helped to improve the essential amino acids contents in the seed storage proteins of a number of crop plants. *E.g.* overproduction of lysine by de-regulation. The four essential amino acids namely lysine, methionine, threonine, and isoleucine are produced from a non-essential amino acid aspartic acid. The formation of lysine is regulated by feed back inhibition of the enzymes aspartokinase (AK) and dihydrodipicolinate synthase (DHDPS). The lysine feedback- insensitive genes encoding the enzymes AK and DHDPS have been respectively isolated from E. Coli and Cornynebacterium. After doing appropriate genetic manipulations, these genes were introduced into soybean and canola plants. The transgenic plants so produced had high quantities of lysine.

Diagnostic and Therapeutic Proteins: Experiments are going on to use transgenic plants in diagnostics for detecting human diseases and therapeutics for curing human and animal diseases. Several metabolites and compounds are already being produced in transgenic plants *e.g.* the monoclonal antibodies, blood plasma proteins, peptide hormones, cytokinins etc. The use of plants for commercial production of antibodies, referred to as plantbodies, is a novel approach in biotechnology. The first successful production of a functional antibody, namely a mouse immunoglobulin IgGI in plants, was reported in 1989. This was achieved by developing two transgenic tobacco plants-one synthesizing heavy chain gamma- chain and other light kappa- chain, and crossing them to generate progeny that can produce an assembled functional antibody. In 1992, C.J. Amtzen and co-workers expressed hepatitis B surface antigen in tobacco to produce immunologically active ingredients via genetic engineering of plants.

Several other therapeutic proteins have also been produced like haemoglobin and erythropoietin in tobacco plants, lactoferrin in potato, trypsin inhivitor in maize etc. The first proteins/enzymes that were produced in transgenic plants (maize) are avidin and beta-glucuronidase and are used in diagnostic kits.

EDIBLE VACCINES IN CROP PLANTS

Crop plants offer cost-effective bioreactors to express antigens which can be used as edible vaccines. The approach is to isolate genes encoding antigenic proteins from the pathogens and then expressing them in plants. Such transgenic plants or their tissues producing antigens can be eaten for vaccination/immunization (edible vaccines).

The expression of such antigenic proteins in crops like banana and tomato are useful for immunization of humans since banana and tomato fruits can be eaten raw. Transgenic plants (tomato, potato) have been developed for expressing antigens derived from animal viruses *e.g.* rabies virus, herpes virus. In 1990, the first report of the production of edible vaccine (a surface protein from Streptococcus) in tobacco at 0.02 per cent of total leaf protein level was published in the form of a patent application under the International Patent Cooperation Treaty.

The first clinical trials in humans, using a plant derived vaccine were conducted in 1997 and were met with limited success. This involved the ingestion of transgenic potatoes with a toxin of E. coli causing diarrhea. The process of making of edible vaccines involves the incorporation of a plasmid carrying the antigen gene and an antibiotic resistance gene, into the bacterial cells *e.g.* Agrobacterium tumefaciens. The small pieces of potato leaves are exposed to an antibiotic which can kill the cells that lack the new genes. The surviving cells with altered genes multiply and form a callus. This callus is allowed to grow and subsequently transferred to soil to form a complete plant. In about a few weeks, the plants bear potatoes with antigen vaccines. The bacteria E.coli, V. cholerae cause acute watery diarrhea by colonizing the small intestine and by producing toxins.

Chloera toxin (CT) is very similar to E.Coli toxin. The CT has two subunits, A and B. Attempt was made to produce edible vaccine by expressing heat labile enterotoxin (CT-B) in tobacco and potato. Another strategy adopted to produce a plant-based vaccine, is to infect the plants with recombinant virus carrying the desired antigen that is fused to viral coat protein. The infected plants are reported to produce the desired fusion protein in large amounts in a short duration. The technique involves either placing the gene downstream a subgenomic promoter, or fusing the gene with capsid protein that coats the virus.

Advantages of Edible Vaccines: The edible vaccines produced in transgenic plants will sole the storage problems, will ensure easy delivery system by feeding and will have low cost as compared to the recombinant vaccines produced by bacterial fermentation. Vaccinating people against dreadful diseases like cholera and hepatitis B, by feeding them banana, tomato, and vaccinating animals against important diseases will be an interesting development.

Biodegradable Plastics: Polythenes and plastics are one of the major environmental hazards. Efforts are on to explore the possibility of using transgenic plants for biodegradable plastics. Transgenic plants can be used as factories to produce biodegradable plastics like polyhydroxy butyrate or PHB. Genetically engineered *Arabidopisis* plants can produce PHB globules exclusively in their chloroplasts without effecting plant growth and development. The large-scale production of PHB can easily be achieved in plants like *Populus*, where PHB can be extracted from leaves.

Molecular Breeding: The term molecular breeding is frequently used to represent the breeding methods that are coupled with genetic engineering techniques. Up till now, conventional breeding methods have been used to meet the food demands of the growing world population and the challenges of poverty and improved crop production and yields. However in the years to come, the development in the agriculture yields and techniques is going to be due to the use of molecular breeding programme.

Linkage analysis which deals with the studies to correlate the link between the molecular marker and a desired trait is an important aspect of molecular breeding programme. In the past, linkage analysis was carried out by use of isoenzymes and the associated polymorphisms. Now a days, molecular markers are being used. Molecular breeding involves breeding using molecular (nucleic acid) markers.

A molecular marker is a DNA sequence in the genome which can be located and identified therefore molecular markers can be used to identify particular locations in the genome. Due to mutations, insertions, deletions, etc. the base composition at a particular location may be different in different plants. These differences, termed polymorphisms, allow DNA markers to be mapped in a genetic linkage group.

Generally, there are three types of markers used in screening/selection:

- Morphological marker based on visible character (phenotypic expression) *e.g.* flower colour, seed colour, height, leaf shapes, etc. Morphological markers could be dominant or recessive. There are certain constraints in using these markers as the morphological markers are easily influenced by environmental factors and thus may not represent the desired genetic variation. Some of the visible markers have not much role to play in the plant breeding programme.
- Biochemical marker: The proteins produced by gene expression are also used as markers in plant breeding programmes. The most commonly used are isozymes, the different molecular forms of the same enzyme. Each individual variety has its own isozyme variability (profiles) which can be detected by electrophoresis on starch gel.
- Molecular marker based on DNA polymorphism detected by DNA probes or amplified products of PCR, e.g.Restriction fragment length polymorphism (RFLP), Randomly Amplified polymorphic DNA (RAPD), variable Number Tandom Repeats (VNTR), Microsatellites, etc. Plant breeders always prefer to detect the gene as molecular marker, although it is not always possible. Molecular markers provide a true representation of the genetic make up at the DNA level. They are consistent and free from environmental factors, and can be detected much before the development of plants occur.

The advantage with a molecular marker is that a plant breeder can select a suitable marker for the desired trait which can be detected well in advance. A large number of markers can be generated as per the needs.

The molecular markers to be used in plant breeding programme should have the following characteristics:

- The marker should be closely linked with the desired trait,
- The marker screening methods should be effective, efficient, reproducible and easy to carry out,
- The entire analysis should be cost effective.

Molecular makers are of two types:

- *Molecular makers based on nucleic acid (DNA) hybridization*: This involves the cloning of the DNA piece followed by the hybridization with the genomic DNA, which is later detected. The Restriction fragment length polymorphism (RFLP) was the very first technology employed for the detection of polymorphism, based on the DNA sequence differences. RFLP is mainly based on the altered restriction enzyme sites, as a result of mutations and recombinations of genomic DNA. The procedure involves the isolation of genomic DNA and it's digestion by restriction enzymes. The fragments are separated by electrophoresis and finally hybridized by incubating with cloned and labeled probes.

- *Molecular markers based on PCR amplification*: Polymerase chain reaction (PCR) is a novel technique for the amplification of selected regions of DNA. The most important advantage is that even a minute quantity of DNA can be amplified and the PCR- based molecular markers require only a small quantity of DNA to start with. Random amplified polymorphic DNA (RAPD) markers use PCR amplification where the DNA is isolated from the genome and is denatured. The template molecules are annealed with primers and amplified by PCR. The amplified products are separated on electrophoresis and identified. Based on the nucleotide alterations in the genome, the polymorphisms of amplified DNA sequences differ which can be identified as bends on gel electrophoresis.

Amplified fragment length polymorphism (AFLP) is a novel technique involving a combination of RFLP and RAPD. AFLP is based on the principle of generation of DNA fragments using restriction enzymes and oligonucleotide adaptors (or linkers), and their amplification by PCR.

MICROSATELLITES

Microsatellites are the tandemly repeated multiple copies of mono-, di-, tri-, and tetra nucleotide motifs. In some instances, there are unique flanking sequences present in the repeat sequences. Primers are designed for such flanking sequences to detect the sequence tagged microsatellites (STMS) which is done by PCR.

Commercial use of Transgenic Plants: The main goal of producing transgenic plants is to increase the productivity. In 1995-96, transgenic potato and cotton plants were used commercially for the first time in USA. By the year 1998-99, five other major transgenic crops cotton, maize, canola, soybean, and potato were introduced to the farmers.

These accounted for about 75 per cent of the total area planted by crops in USA. There are still a lot of concerns regarding the harmful environmental and hazardous health effects of transgenic plants.

The major areas of public concern are- the development of resistance genes in insects, generation of a super weeds by mutation etc. Certain other legal and regulatory hurdles pertaining to commercial use of transgenic plants, needs to be addressed.

Bioethics in Plant Genetic Engineering: There are issues and concerns regarding the use of transgenic crops and their effects on the health and the environment in general.

The major concerns about GM crops and GM foods are:

- *Effect of GM crops on biodiversity and environment*: As the GM crops are created artificially, there is no natural process of evolution in their development. Hence, there is a question of this affecting the biodiversity and overall effect on the environment.
- *The risk of transfer of transgene from GM crops to pathogenic microbes*: Antibiotic marker genes are used to identify and select the modified cells. If GM food containing antibiotic resistance marker gene is consumed by animals and humans, there is a risk that the transgene will transfer from GM food to microflora of human and animals. This may lead to the gut microbes to become resistant to antibiotics.
- The transfer of genes from animals into Gm crops for molecular farming may change the fundamental vegetable nature of plants.
- The GM crops may bring about changes in evolutionary patterns. The plants adapt to the changing environment in the natural way by changing their genes and developing better races with superior traits which ultimately leads to the development of evolved races and varieties. What will be the evolutionary pattern of the GM crops? There are concerns about the effect of transgene flow from GM crops to other non-GM plants and the alteration of these non-GM crops.
- There is a risk of transferring allergens (usually glycoproteins) from GM food to human and animals.
- There is a risk of "gene pollution" *i.e.* transfer of transgene of GM crop through pollen grains to related plant species and development of super weeds.
- There are also some religious issues related to the consumption of transgenic plants with animal genes introduced into them, especially, for some strict vegetarian people and some ethnic groups with certain food preferences and restrictions.
- There is a need to study thoroughly as to how the genetically engineered plants will affect the ecological balance, once they are released in the environment.

GENE CLONING OF PLANT TRAITS

To the average consumer, plant propagation might simply mean growing a new plant from a seed. Experienced gardeners and farmers, however, know that they are numerous ways to propagate, both sexual and asexual. The asexual methods usually create an exact genetic duplicate, or clone of the parent plant. This is usually highly desirable when a gardener finds a plant with superior genetic traits. Plant cloning is almost as old as agriculture itself, and some species of trees originated as plants that were cloned on a mass scale. Plant gene cloning is currently limited to simply inherited single gene traits. The transfer of genes to plants is limited to single genes or, at most, two to three genes encoding a particular, well-characterized biochemical pathway. Manipulation of multigenic traits is currently only feasible within the existing gene pool, sometimes made more efficient by markerassisted selection. When traits are regarded in the molecular genetic sense, often the coding sequence is the major consideration.

Obviously, the coding sequence is of major importance as this is the template for determining the amino acid sequence of the final protein or enzyme which determines the phenotype. However, a major factor in controlling the phenotype is the presence of regulatory sequences surrounding the coding sequence in the form of upstream (promoters and enhancers) and downstream (terminator) regulatory sequences and, in many eukaryotic genes, introns. When traits are being manipulated via genetic engineering, the role of these regulatory sequences cannot be trivialized.

Put simply, these sequences control such things as the level of gene expression, tissue/organ specificity, developmental expression and response to particular abiotic/biotic stimuli. The regulatory sequences that have been subjected to greatest attention in plants are the upstream promoters.

For proper expression of a seed storage protein, for example, the gene of interest must be under the control of a promoter which expresses in developing endosperm (monocots) or cotyledons (dicots). Expression of other genes may be required only in anthers, root hairs or photosynthetic tissues, or in response to a wound such as that caused by a chewing insect or penetration of a fungal infection structure. In the earliest cases of transgenic plants, introduced genes were placed under the control of constitutive promoters of bacterial (octopine or nopaline synthetase) or viral (cauliflower mosaic virus (CaMV) 35S) origin.

These promoters have been found to give adequate expression of selectable marker and reporter genes in dicotyledonous species. However, expression is commonly not of a sufficient level in monocots, especially cereals, for good selection with antibiotic or herbicide resistance genes. And, as already discussed, such a constitutive pattern of gene expression is commonly not desirable for most traits.

For cereals, a number of other promoters have been developed to achieve high-level constitutive expression of selectable markers or reporter genes. These include sequences cloned directly from cereal tissues such as the ubiquitin and rice actin 1 promoters and synthetic promoters involving various spliced sequences from a range of sources such as the Emu promoter. However, to properly control gene expression such that the introduced character is most efficiently targeted without other pleiotropic effects, it is generally found that regulatory sequences of plant origin are desirable.

Further to this, expression is usually best controlled when a cereal promoter is used in a cereal or a dicot promoter in a dicot. The most widely used means of cloning such regulatory sequences is by using cDNA libraries from specific tissue types and looking for abundant signals. Naturally, cDNA libraries will not contain the regulatory sequences, hence the abundant cDNA clones must then be used to screen a genomic library so that the regulatory sequences can be identified and cloned.

METHODS FOR CLONING PLANT TRAITS

Many of the first plant traits to be cloned were those in which some degree of biochemical understanding preceded any work with DNA. The understanding may have come from plant systems or from outside the plant kingdom.

Most of the cloned genes in these cases were enzymes which were part of a well-characterized biochemical pathway or had a protein structure which was otherwise well known, such as seed storage proteins. Cloning genes based on a knowledge of the protein sequence can be achieved by working backwards from the gene product (amino acid sequence) to the coding sequence or the DNA template.

Such an approach has been used for cloning many seed storage proteins, such as the 2S sulphur- rich proteins from Brazil nut. Genes of plant origin have been used in transgenic plants to improve resistance to insect pests. One of the first to be characterized was the cowpea trypsin inhibitor, a protease inhibitor which acts as an antifeedant to a range of insect pests. The protein is naturally produced in seeds of some genotypes of cowpea (*Vigna unguiculata*), which was shown to confer lepidopteran insect resistance when expressed constitutively in tobacco.

Another such report from the same group at Durham University described the use of a snowdrop lectin gene to confer aphid resistance in transgenic tobacco, the first report of genetically engineered resistance to a sucking insect in plants. The lectin from snowdrop (*Galanthus nivalis*), known as GNA, has been demonstrated to be effective against two major rice pests, the brown planthopper and the green leafhopper, as well as the peach potato aphid. When expressed at 0.1 per cent of total leaf protein in tobacco, GNA is antimetabolic to homopteran pests.

These levels of protein were accumulated in leaf tissues when placed under the control of the CaMV 35S promoter. However, Hilder *et al.* (1995) postulate that better control would be achieved with the use of a phloem-specific promoter, such as the sucrose synthase 1 promoter from rice. Another significant group of genes which have been cloned based on prior knowledge of the end product are the seed storage genes, including both seed storage proteins and lipids.

In both cases there was considerable biochemical knowledge of the pathways involved in biosynthesis and storage. For seed storage proteins, the genes can be cloned based on the knowledge of the amino acid sequence of the mature protein and targeting signal or transit peptide. Molecular genetic manipulation of vegetable oils is a major area of research in the developed world.

While approximately 90 per cent of vegetable oils are used for human consumption, there are emerging markets for industrial applications including lubricants, paints, cosmetics, plasticizers and soaps. With increasing emphasis on quality and dietary considerations, there is a desire in developed countries to substantially alter the types and balances of fatty acids found in the major oil crops. Six crop species (soybean, oil palm, rapeseed/canola, sunflower, cottonseed and peanut) account for 84 per cent of the world's vegetable oil production, with the major fatty acids produced being palmitic, stearic, linoleic, linolenic and oleic acids. However, almost 200 different fatty acids are produced by plants, mostly occurring in non-domesticated plant species.

Hence, genetic manipulation of the major crop species, either via conventional breeding or genetic engineering, has the potential to modify the fatty acid types produced and stored by seeds such as soybean and canola. The biochemistry of fatty acid biosynthesis is well established for many of the major pathways, with enzymes identified which control the chain length of fatty acids, and the degree of saturation, or number of double bonds in the fatty acid chain.

The oils stored in seeds have been altered by downregulation of particular enzymes using antisense RNA techniques. The oleic acid content of rapeseed oil has been increased to 83 per cent (from 62 per cent in conventionally bred cultivars) by downregulating the enzyme D12-oleate desaturase using antisense RNA. A similar result has been achieved using co-suppression, which has produced rapeseed oil with 87 per cent oleic acid content.

Genetic engineering approaches have also been used to introduce totally novel fatty acids to major oilseed crops. Petroselinic acid, an isomer of oleic acid, is found in abundance in the spice plant, coriander (*Coriandrum sativum*). This oil has uses in cosmetics and pharmaceuticals, and its oxidation leads to the formation of lauric acid, used in detergents, and adipic acid, which is used in nylon production. Hence this is a potentially valuable oil which may enable soybean and canola crops – *i.e.* renewable resources – to replace some of the nonrenewable carbon fuels which are currently used for many industrial polymer and lubrication applications. A cDNA clone from coriander encoding an acyl-ACP desaturase has been genetically engineered into tobacco, resulting in the accumulation of petroselinic acid. Oil crops such as rapeseed and soybean could be genetically modified to produce this fatty acid on a commercial scale. Hence, here is an example whereby a minor crop plant has become a donor of genetic diversity which will enable the commercial production of a very useful array of products for food and industrial purposes.

Cloning Genes Known Only by Phenotype: However, a more challenging task has been to move beyond proteins and pathways about which sufficient biochemistry is known into cloning genes known by phenotype only. By far the majority of genes fit into this category, which includes many of the most important plant traits such as disease resistance, adaptation to abiotic stresses, phenology and flowering behaviour and fertility. Only in the past few years have genes for traits known only by phenotype been cloned. One of the most interesting areas of plant biology is the understanding of plant–pathogen interactions at the molecular level.

The major methods used for cloning genes known only by phenotype are:

- Transposon tagging.
- Map-based or positional cloning.
- T-DNA tagging.

There have been a number of plant genes cloned using all these methods or modifications of them. There are also examples where a combination of methods has been used to expedite the rate of identification of the unknown sequence.

Transposon Tagging: The proposal that specific pieces of DNA are able to autonomously excise from a particular locus and insert, theoretically at random, into another linked or non-linked locus was first proposed by Barbara McClintock. Transposable elements are now accepted as ubiquitous, and are particularly well characterized in organisms such as *Escherichia coli*, *Drosophila* and maize. In two plant species, maize and the snapdragon (*Antirrhinum majus*), transposable elements are sufficiently well characterized to be used for mutagenesis and gene tagging.

The best understood system is the maize *Ac/Ds* (*Activator/Dissociation*) family of elements. The behaviour of the *Ac* element is such that it can excise autonomously from its locus, under the control of its own transposase, and insert into another chromosomal position. Where the *Ac* element is in or near a gene, that gene is inactivated (*i.e.* not expressed), and will appear phenotypically as a recessive mutation. When *Ac* transposes (excises and reinserts at another locus), the gene is commonly returned to normal function, and hence is termed a revertant. Should this *Ac* element then insert into another gene (transcribed or regulatory sequence), then a new recessive mutation will manifest.

Transposition can occur either germinally, which results in an altered phenotype for the whole plant, or somatically, which results in a mosaic mutation, and can often be detected as a sectoring of plant or seed colour. The *Ds* element is actually a group of elements derived from defective *Ac* elements, with a common feature being their lack of effective *Ac* transposase. Hence these elements are non-autonomous, and will only transpose in the presence of an *Ac* element capable of producing transposase. Equipped with sequence and map information of an *Ac* element in maize, geneticists have been able to clone important maize genes using the transposon-tagging approach.

Table. Plant Transposable Elements which have been used for Transposon-Taggine

Element	Species	Type of Element	Size
Ac	Maize	Autonomous	4.6
Ds	Maize	Non-autonomous	0.5–30
Tam1	Snapdragon	Autonomous	15
Tam2	Snapdragon	Non-autonomous	5
En/Spm	Maize	Autonomous	8.3

When a gene is inactivated by an *Ac* element, proof of which can be demonstrated by a low frequency of trait reversion, the *Ac* element can be used as a probe for 'fishing-out' the tagged gene. A tagged gene can be cloned by performing Southern blot analysis with the *Ac* element used as probe, and any polymorphism represents an excision/insertion event.

The technique is described more fully in the review by Walbot. The *Ac* system has proven extremely useful for cloning maize genes for traits such as plant pigment genes (*bronze*; the anthocyanin pathway), starch biosynthesis (*opaque2*), and abscisic acid (AbA) insensitivity.

Other maize transposable elements such as *Mutator* (*Mu*) and the *En/Spm* system have been used to clone maize genes, as has the *Tam* system in snapdragon. As a result, the search began for transposable element systems in other species so that a similar approach could be taken. However, it was soon demonstrated that the well-characterized *Ac/Ds* system could be genetically engineered into other plant species including tobacco, *Arabidopsis*, tomato and petunia.

It was subsequently shown that the *Ac* element acted in a similar manner in these heterologous species, undergoing germinal and somatic excision and insertion, and maintaining the propensity to reinsert at a linked site (usually within 4 cM). As a result, the frequency of mutation in the surrounding area approaches 1024.

It was not until 1993 that a plant gene was cloned using the heterologous system, that of the *Ph6* gene in petunia, which is involved in floral pigmentation. Using the transposon-tagging approach, a number of plant disease resistance genes (R genes) have now been cloned. The first of these was the tomato *Cf-9* gene, which confers resistance to race 9 of the fungal pathogen *Cladosporium fulvum*, causal agent of leaf mould. The gene was tagged using a modified *Ac/Ds* system. The system relied on the use of a stabilized *Ac* element (s*Ac*), which was effectively a disabled *Ac* element that nevertheless produced *Ac* transposase and hence enables *Ds* transposition.

The two elements are maintained in separate lines, and hence are both stable until controlled crossing is performed. This gives a significant advantage in that *Ds* insertion mutants can be maintained indefinitely in the homozygous or hemizygous form, and revertants can be isolated by crossing lines with s*Ac* tomatoes. The transposon-tagging approach used to isolate the *Cf-9* gene, which was known by phenotype only, did have considerable advantages in that much was known of the pathogen. The hypersensitive response elicited by the virulent strain fitted into the gene-for-gene hypersensitive response proposed by Flor, with a dominant avirulence gene (*Avr9*) interacting with the *Cf-9* gene to form an incompatible or hypersensitive response.

*Avr*9 had previously been cloned and fully characterized, and was known to specify a 28 amino acid peptide which alone could elicit a necrotic response in resistant tomatoes. Crosses were made to develop a transgenic tomato line homozygous for *Cf-9*, but heterozygous for *Ds*.

As can be seen, most seedlings died from systemic hypersensitive response, with *Avr9* and *Cf-9* present in all cells. The only survivors would be individuals where *Ds* had inserted into the *Cf-9* gene. Where this happened in plants carrying s*Ac*, the seedlings were variegated for the hypersensitive response (somatic excision), and in the absence of s*Ac*, the

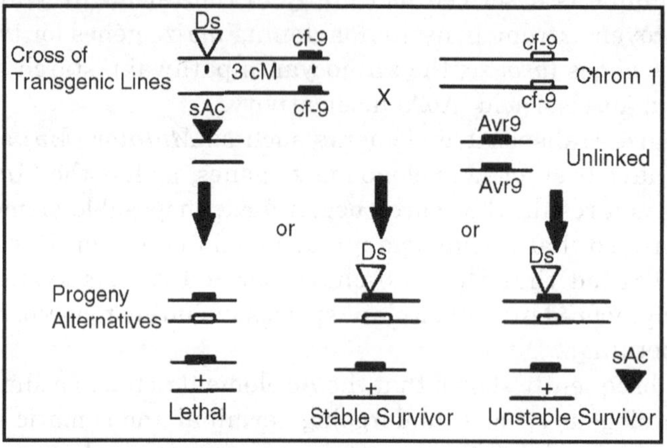

Fig. Graphical Representation of the Crosing Strategy of Transgenic Ac/Ds Tomato Lines.

This work was a major undertaking, involving the germination of approximately 160 000 progeny, yielding 118 survivors, representing 63 independent mutations. After considerable effort to characterize these, the *Cf-9* gene was cloned and characterized, with an open reading frame encoding an 863 amino acid protein.

Map-based Cloning: The previously described transposon-tagging scheme utilized to clone *Cf- 9* had an element of the map-based approach by utilizing the phenomenon of *Ac/Ds* transposition to linked sites, with a *Ds* element within 3 cM of the locus of interest, which greatly enhanced the likelihood of insertional mutagenesis. However, map-based cloning requires a much tighter genetic linkage (in centimorgans or recombination fraction), which in turn needs to be converted into a physical distance (kbp of DNA sequence). The map-based, or positional cloning concept relies on building a well-saturated genetic map of DNA sequences of the organism in question.

The genetic map is built up in a polymorphic mapping population, commonly an F2, backcross (BC) or recombinant inbred, with markers such as RFLPs (restriction fragment length polymorphisms). The first RFLP map to be made was of human. Based on such maps, important human genes responsible for disorders such as cystic fibrosis and muscular dystrophy have been cloned in this manner. Genetic maps of plant species have been made with DNA markers in virtually all important plant species, particularly the major cereals, grain legumes, oilseeds and important vegetable crops including tomato and potato. The RFLP marker system is based on Southern blot analysis, while many of the newer technologies use the polymerase chain reaction (PCR). These include random amplified polymorphic DNA (RAPDs), amplified fragment length polymorphisms (AFLPs), and a number of different marker systems based on simple sequence repeats (SSRs), which may use PCR, such as inter simple sequence repeat (ISSR)-PCR, or Southern analysis.

The applications of DNA markers to plant genetic analysis and breeding are many, and the subject has been reviewed by various authors. The population for saturation mapping should be segregating in a Mendelian manner for the trait of interest.

This requires a lot of tedious but demanding work, and it may be that particular populations such as near isogenic lines, or DNA pooling strategies, such as bulked segregant analysis, will allow the mapping effort to be more targeted to the region of interest, hence improving the rate of map saturation. It is desirable to obtain extremely close genetic linkage to the locus of interest (<1 cM), with flanking markers if possible. It can be of considerable advantage to identify a marker which co-segregates (say within 0.1 cM) with the trait.

The next step is to convert the genetic map into a physical map of the chromosome region, translating distances from centimorgans to nucleotides. The Gene of Interest (in this illustration,R) is Flanked by Markers such as RFLPs or SSRs. A Series of Overlapping YAC Contigs is used to cover th Chromosomal Region Containing the Gene.The YAC Which Contains the Complete Gene is Marked in Bold.This is then Subcloned or Matched to a cDNA Library to Idenfify the Gene(s) Contained the Region, and Ultimately Identify the R Gene.

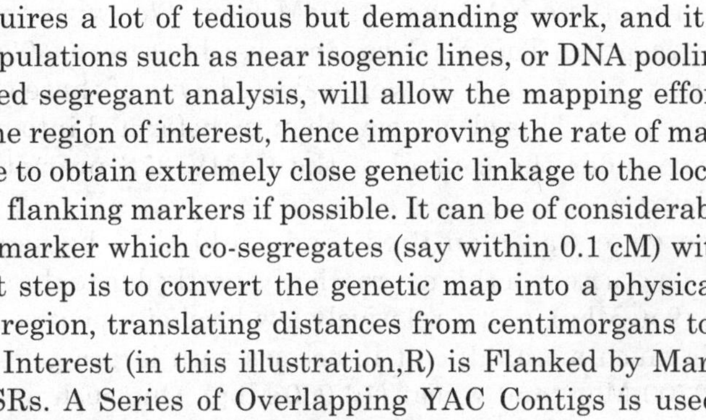

Fig. Generalized Strategy of Chromosome Walking Used in Map-based gene Cloning.

Physical mapping then involves the use of YAC or BAC clones which span the region of interest. These YAC/BAC contigs will span the region between two flanking DNA marker sequences as shown in Figure, a process known as chromosome walking. In this example, two YAC/BAC clones contain the gene of interest.

Methodology is being developed to enable BAC clones to be directly transformed into plants, to find which one contains the gene of interest (in the example, the *R* gene). Alternatively, the YAC/BAC must be subcloned (for example as a cDNA library), and these subclones can be put into the susceptible plant and tested for the R phenotype. The first R gene to be cloned in plants via map-based cloning was the *Pto* gene, which confers resistance to

Pseudomonas syringae pv. *lycopersici*, the causal agent of bacterial speck in tomato.

The starting points for the work were:

- A high-density RFLP map of tomato.
- A tomato YAC library.

The *Pto* locus had been previously mapped to tomato chromosome 5 and one RFLP marker co-segregated with *Pto*. Hence this was a useful starting point to screen the YAC library. A 400 kb YAC was identified, and via inverse PCR, some end-specific probes were generated.

The left arm of the YAC was 1.8 cM from *Pto*, with the right arm perfectly co-segregating. This YAC was then used to screen a 920 000 plaque leaf cDNA library, from which 30 were selected. One of these was shown to co-segregate with *Pto*. This cDNA was then transformed into a susceptible tomato cultivar under the control of the CaMV 35S promoter, where it conferred resistance.

Hence the gene was identified, and could then be sequenced and characterized. The success of this approach was partly due to the small genome size of tomato (950 Mbp), and a relatively low level of repeat sequence. Such an approach is much more difficult in species with larger genomes such as maize (2300–2700 Mbp) and wheat (16 000 Mbp). One such example of this problem is the 79 000 clone maize YAC library produced by Zeneca Seeds.

As reported by K. Edwards, only 15 per cent of the YACs have unique ends, with the remainder terminating in repeat sequences, which makes chromosome walking very difficult. The synteny of genomes is a major advantage in this instance. It is now well established that there is a high degree of synteny of genome organization, as seen in the co-linearity of maps of the most important grass genomes.

This has resulted in the great interest in using this co-linearity to facilitate cloning of genes in large genomes such as maize and wheat, by starting with the smaller genomes, particularly rice, and to a lesser extent, sorghum. One particular study of maize YACs demonstrated that a single YAC with the *Adh1* sequence contained 36 different repetitive sequences, whereas these repetitive sequences were largely absent in corresponding areas of the rice and sorghum genomes, a pattern seen also for the *a1–sh2* region in all three genomes. In a similar manner, many genes of interest for *Brassica* improvement are being identified in the smaller *Arabidopsis thaliana* genome. Paterson *et al.* have recently demonstrated the potential of this approach, by identifying quantitative trait loci (QTLs) for seed mass and phenology which were co-linear on molecular marker maps of maize, sorghum and rice.

They proposed that, as the maize genome is four and six times larger than the sorghum and rice genomes, respectively, the genes of interest in maize may be cloned in sorghum/rice more expediently. It must be remembered that QTLs are subject to genotype 3 environment interaction, which will be a major complicating factor in attempts to clone and characterize a QTL.

T-DNA Tagging: When T-DNA is inserted into the plant genome, it does so via illegitimate recombination. In the same way as a transposon, the insertion of T-DNA into a gene will inactivate it, which phenotypically will appear as a mutation.

The comparative advantage of such a method is that stable mutations will be conferred. This can be a disadvantage, however, as reversions, such as those made possible by the *Ac/Ds* system are not possible.

It is also the case that each independent transformation event will result in just one mutation event (or more if multiple independent insertions are made), whereas the heterologous transposon approach allows multiple mutation events to take place from a single initial transformant. Hence, mutated genes with T-DNA insertions can be cloned, in much the same manner as those with transposon insertions, by using part of the T-DNA as a probe in Southern blots. The procedure for selecting out the tagged gene is quite straightforward, particularly when the plant contains a single T-DNA copy.

However, this means that a large number of independent transformants must be generated. This can be achieved with species such as *Arabidopsis*, where Feldmann demonstrated that a seed-based transformation system using *Agrobacterium* can be quite efficient. Feldmann demonstrated that by inoculating approximately 1000 seeds, approximately 300 000 selfed T2 seed can be collected. They estimated that approximately 8000 transformants could be recovered from an experiment of this scale and, using this approach, cloned a gene involved in trichome development and a dwarf gene from *Arabidopsis*. A regulatory gene in the ethylene biosynthesis pathway of *Arabidopsis* has also been cloned using the T-DNA tagging strategy. With modification, T-DNA tagging methodology can be used to clone plant promoter sequences.

Koncz *et al.* designed a vector which contained a promoterless antibiotic selection gene (aminoglycoside phosphotransferase II) and transformed this into *Nicotiana* and *Arabidopsis*. In this way, promoters could be 'trapped' as plants that were regenerated on selective antibiotic were antibiotic resistant because the T-DNA was inserted downstream of a promoter such that the gene was expressed. The most serious limitation to the T-DNA tagging approach is that large numbers of independent transformation events must be generated, which is not currently possible with most species.

MOLECULAR MARKERS IN GENE TRANSFER

Once useful genes have been located on a framework map, it is then possible to use the molecular markers to facilitate the transfer of the useful genes between species and strains in an efficient manner. As stated earlier, recombination is not a frequent event along a chromosome and, furthermore, two recombination events rarely occur within a distance of 15 to 20 cM of each other.

It therefore follows that if a useful target gene (a major gene or a QTL) is known to be located within a region of chromosome flanked by two markers that are no more than 15 cM apart, then these markers can be used to follow and control the progression of the target gene through successive stages of a breeding programme. Clearly, the alleles at the marker loci have to differ in the donor and recipient cultivars, but any chromosome that contains A1 and B1 at the marker loci will also contain T1, the target allele to be transferred. Should single recombination occur close to T, then the chromosome will contain A1 or B1 but not both.

Providing the markers are less than 15 cM apart double recombination resulting in A1T2B1, and hence loss of T1, can be safely ignored. If the location of T is not accurately known, then three or more marker loci may be needed to be sure of safely bracketing the region containing T without fear of double recombination.

There are conflicting requirements in marker-assisted gene transfer: the need to keep the bracketed region large enough to be sure of holding the gene to be transferred whilst not having it so large that too many other linked but undesirable alleles are transferred with it. Let us now consider the basic procedure of gene transfer.

Traditionally, breeders have introduced a useful allele into their cultivars by backcrossing the original F1 to the commercial cultivar whilst selecting at each generation for the desired allele. As was shown earlier, this can result in a very large region of chromosome around the target gene surviving even after ten generations of backcrossing – a phenomenon known as linkage drag. With markers, however, it is possible to reduce linkage drag considerably whilst simultaneously reducing the number of backcrossing generations to two or three instead of the normal six to eight. Moreover, the amount of plant material raised at each generation can be reduced because much of the molecular genotyping can be achieved at the juvenile stage.

The aim is to hold the target allele, T1, heterozygous throughout the backcrossing process, by selecting for the flanking markers, whilst simultaneously selecting for the genotype of the recurrent, commercial cultivar at all other loci. The latter can readily be achieved by having as few as four or five well-spaced markers on all chromosomes and selecting for recurrent parent alleles at each generation. Depending on the chromosome number, some individuals in the first backcross generation will be homozygous for several whole chromosomes whilst still being heterozygous for the target gene.

Those which have most of the recurrent parent chromosomes are selected and backcrossed again, and if necessary the process is repeated for a third generation. For each generation, fewer markers need to be screened as those found to be homozygous in the previous backcross no longer have to be checked.

Unless the position of the target gene is very accurately defined, it is prudent to have several markers covering its likely position, again with the proviso that they need to be separated by no more than 15 cM.

At the final stage of the introgression process, a backcross individual which is homozygous for most of the recurrent parent alleles is selfed or, if this is not possible, intercrossed with a similar genotype. From among the progeny, individuals are chosen which have different combinations of markers bracketing the target region.

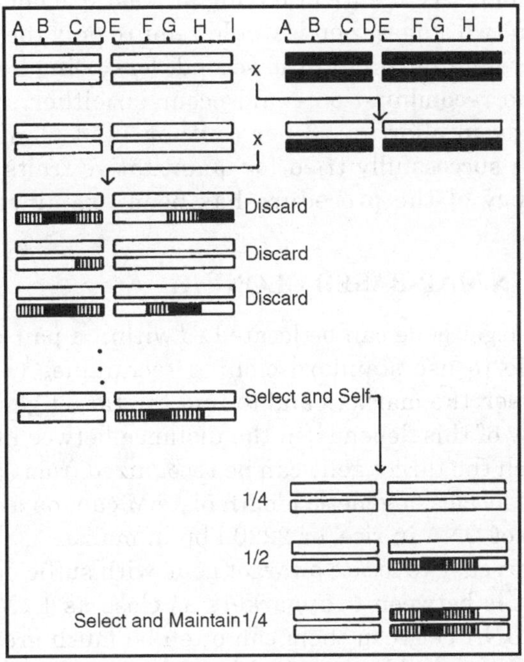

Fig. The use of Molecular Markers to Facilitate Gene Introgression.

These can be selfed again and homozygotes for the different sequences identified and multiplied for reassessment for the target trait. Some will fail to show the target trait because of recombination, but of those that do, the lines with the shortest sequence of donor markers are selected for further trials.

The approach can obviously be adapted to the simultaneous introgression of several genes: no new principles are involved but the screening process becomes more elaborate. It is always possible at a later stage to reduce the length of the introgressed region around T by looking at other, more closely linked, markers.

Similarly, it is wise to check the origins of the other chromosomes to avoid having introgressed unwanted regions through missing double recombinants in the central regions of individual chromosomes or single recombinants towards their ends. If such errors are found they can easily be corrected by another round of backcrossing to the recurrent parent. When attempting to reduce the length of donor chromosome surrounding the target gene, it is important to realise that this should be done in two successive generations.

If the sites of the required recombination events are less than 15 cM apart, then the simultaneous double event will not occur. Even if they are far enough apart for the double event to occur, the probability of such an event may be too low to be practicable. If two recombination events each have a probability of 0.05 (*i.e.* 5 per cent), their combined probability is at best 0.0025.

Therefore, it would be necessary to genotype approximately 1200 individuals to be 95 per cent sure of having at least one double recombinant. If it was tackled in two stages then it would require only the genotyping of 86 (= 28 in the first round plus 58 in the second; less being needed in the first round because the recombination could occur on either side of T), a very considerable saving in effort of 93 per cent, or 1114 plants! Marker-aided selection has been successfully tried for quantitative traits in several crops while the efficiency of the procedure has been examined by Lande and Thompson (1990).

TARGET GENE IN MAP-BASED CLONING

Providing a target gene can be located to within a pair of marker loci, it should be possible to use standard cloning techniques to walk along the chromosome between the markers and to find the target gene.

The feasibility of this depends on the distance between the markers and the ease with which the target gene can be recognized from among the clones. Depending on the species, a genetic length of 1 cM can, on average, consist of as few as 280 kbp of DNA in rice, or 2220 kbp in maize.

It is not always easy to locate a target gene with sufficient accuracy to be able to say that it is between two markers as close as 1 cM, and hence the actual lengths of DNA between them can often be much greater. With QTLs one can seldom achieve anything approaching this accuracy by existing methods of conventional mapping. Furthermore, these are average distances per centimorgan; some chromosomal regions could be much longer or, indeed, less.

Increases in mapping reliability can be obtained by using what are called near isogenic lines (NILs), which are lines obtained by selfing or backcrossing and are known to differ from some standard genotype by just a short defined section of chromosome.

Providing this section is delineated by accurately mapped flanking markers, and the NILs which do and do not contain this region can clearly be shown to differ for the target gene, then the target gene has to be in that region. The genetic and physical proximity of the markers can be established by conventional mapping and by gene cloning. Map-based gene cloning involves starting from one of the markers and walking to the next by means of a series of partially overlapping cosmid clones.

Cosmid clones are necessary because they can be up to 50 kb in length and hence require fewer cloning steps in order to cover the DNA between the markers (280 to 2200 kb per cM in the examples cited above). Identifying the target gene is more difficult. Clearly, if the gene is between the two flanking markers it has to be on one of the cloned regions and could possibly be on two overlapping clones. It could be identified by transforming plants with each clone separately and identifying which transformant expresses the effect.

If the gene concerned exhibits classical dominance it would be necessary to transform the homozygous recessive with the dominant allele or the transformant would not be recognized. Alternatively, the gene has to be introduced into a species that does not normally express the gene. Once the effective cloned fragment is identified, the actual location of the gene within the fragment could be identified by successively transforming sub-fractions of the clone or, following sequencing, by looking for potential open reading frames.

In those cases where the target gene produces a known product, cDNA clones derived from tissue likely to be rich in the appropriate mRNA of the target gene can be used to identify likely sequences in the previous overlapping clones. Once the gene has been identified by one of these methods, its structure and activity can be studied in detail. Disease resistance genes are obvious candidates for such studies as their action is very specific and their location on the genome of many species is well documented. However, unless suitable candidate loci.

2

Applications of Gene Biotechnology

Recombinant DNA and cell fusion techniques may soon allow man to change food crops from inside out with great speed and precision, shaping life to fit the environment. The image presented of biotechnology is that the new knowledge of life processes will enable humans to live in a sustainable fashion where economic and ecological efficiency are optimally met.

For this to work, the production sector cannot treat nature as a great and infinite wellspring of resources that provides without being provided for. Have we finally reached a stage in the evolution of technology where new processes of production can be harmonious with natural processes?

This expectation carried in the rhetoric of its promoters set biotechnology apart from other industrial revolutions where the criteria for success almost always explicitly favored economic growth and efficiency over the ideas of sustainable production, biocentrism, and ecological balance between humans and the natural habitat.

Can biotechnology achieve such goals? Does this revolution embody the seeds of a gentler technology? The four chapters of this part of the book explore the applications of applied genetics to the environment and the social response to several key innovations.

Coming as they did in the wake of a new global environmental awareness, the possibilities of releasing genetically engineered organisms into the environment were met with high hopes, skepticism on issues of ecological safety, and some ideological opposition that questioned the wisdom of genetically redesigning plants and animals to increase the human harvest of nature.

THE BACONIAN LEGACY

The desire to control and rearrange species in nature is deeply rooted in the Western scientific tradition. Centuries before genes were understood, scientists contemplated a future in which the natural species were merely starting materials for the conscious rearrangement of the biotic world. For these scientists, the architecture of living forms was likened to the natural minerals of the earth. They were there to be cast into whatever form the human imagination could muster and the life sciences could render. Francis Bacon, the English scientist-philosopher, unleashed his imagination in a 1622 work titled *The New Atlantis* that describes a biological utopia. Bacon's writings heralded a new philosophy of nature and a new epistemology of science.

To have knowledge about the world is tantamount to having power over it. Control over nature was both the goal and the test of knowledge. Bacon provided a literary vision of how scientific knowledge will transform agriculture.

We have also large and various orchards and gardens, wherein we do not so much respect beauty as variety of ground and soil, proper for divers trees and herbs, and some very spacious, where trees and berries are set, whereas we made divers kinds of drinks, besides the vineyards. In these we practice likewise all conclusions of grafting and inoculating, as well of wild trees as fruit-trees, which produce many effects.

And we make by art, in the same orchards and gardens, trees and flowers, to come earlier and later in their seasons, and to come up and bear more speedily than by their natural course they do.

We make them also by art greater much than their nature, and their fruit greater and sweeter, and of differing taste, smell, colour, and figure from their nature. And many of them we so order, as that they become of medicinal use. We have also means to make divers plants rise by mixtures of earths without seeds, and likewise to make divers new plants, differing from the vulgar, and to make one tree or plant turn into another. Bacon's "biotopia" included an experimental programme for redesigning animals.

He described great parks with beasts and birds of all varieties. They were the natural substrate for altering phenotypes to meet human intellectual, practical, or aesthetic needs. Bacon's utopian prophecy is remarkable to the extent that it operated on the same paradigm that exists in the scientific community today, namely, that there are no unalterable properties of animals or plants. He rejected the idea of species barriers or the notion of integrated wholes. Furthermore, he recognized no moral considerations in the remaking of life-forms.

The new secular science was not burdened with the ecclesiastical sanction that humans should not seek to improve upon God's creatures.

JACQUES LOEB AND THE CREATION OF LIFE

In the latter part of the nineteenth century, German-born physiologist Jacques Loeb advanced the field of biological engineering and believed its fundamental task was to produce living matter artificially.

Like Bacon, Loeb refused to accept species boundaries or evolutionary constraints. He actively sought to transcend natural barriers if for nothing more than the challenge of it. Loeb first received international acclaim for his experiments in circumventing biological fertilization.

In 1899, while working with sea urchins, Loeb discovered what he called "artificial parthenogenesis." He devised a method of fertilizing the eggs of sea urchins without male sperm by using inorganic salts. In Philip Pauly's *scientific biography of Loeb*, he notes that "artificial parthenogenesis was a vindication of Loeb's hopes and a model for science to come in which biologists would consciously work to reconstruct the natural order to make it more rational, efficient, and responsive to the ongoing development of engineering science."

Loeb demonstrated that what was believed to be a pure biological function-fertilizing an egg–was reducible to physical chemistry. His discovery brought banner headlines such as "Science Nears the Secret of Life," as well as a host of satirical cartoons. Like Bacon, Loeb rejected the Greek view of science as wisdom and understanding of the world as given.

In its place he substituted conquest and transformation as the goals of science, "utilizing the forces of nature to bring about new combinations, creating things which have never been created outside of nature's workshop." For example, Loeb studied the possibilities of redesigning animals that regenerate certain body parts.

Have succeeded in finding animals in which it is possible to produce at desire a head in place of a foot at the aboral end, without injuring the vitality of the animal. now have animals that have heads on each end of their bodies; thus have animals with bilateral symmetry that in nature have different oral and aboral poles. This idea is now hovering before me that man himself can act as creator even in living nature, forming it eventually according to his will.

Loeb developed a specialized vocabulary to describe his research programme in biological engineering. "Abiogenesis" meant the creation of living things from simple substances, whereas "heteromorphosis" described how a physical process could be used to create biological mutants (two-headed tubularia). Through a series of experiments on lower animals, Loeb sought a "technology of living substance" *(einer Technik der lebenden Wesen.)*

After the late nineteenth century, practical efforts to fulfill Bacon's utopian prophecy were initiated by plant breeders who sought perfection first through the process of breeding, screening, and selection, and subsequently through techniques of cross-fertilization and crop hybridization.

Plant breeders have altered the genetic strains of crops and flowers by methods developed over centuries, and hybrids have been disseminated widely in the environment. The genetic architecture of the hybrid plant is modified by human intervention, but not by human design, since there is considerable randomness and uncertainty to the process. Millions of plants may be screened in search for a few favorable genotypes. Modern genetics goes a step further than its classical counterpart.

No longer is the plant breeder constrained by species boundaries. As Bacon had imagined, plant species may be transformed according to some archetype or blueprint. Seeds, once thought to be the basic units of plants, are displaced from their privileged status by the cellular and subcellular units. A similar change has taken place with respect to animal life. The breeding of animals, which can be traced back to the earliest period of domestication, contributes an artificial form of environmental pressure. Evolution is guided by human intention. Thus a sluggish, overweight sow is the desired phenotype for the feedlot, but not, perhaps, for survival in the wild. Industrial and biological evolution became one and the same. The results are animals that are selected for compatibility with humancentered systems–domestic, agricultural, or food processing. As a method to direct the evolution of animals, breeding is a slow, limited, and ponderous technology in comparison to what may be achieved by the application of molecular genetics to animal reproduction. Since genes can be inserted directly into the fertilized egg, obstacles of species incompatibility are easily overcome.

This was dramatized by several experiments on rodents. The first transgenic experiment on an animal embryo took place in 1981 when a rabbit globin gene was inserted into a mouse. In 1982 rat genes and subsequently human genes were implanted into the fertilized eggs of mice. The result was a mouse twice its normal size.

Also, with the methodology of germline manipulation of animals it is possible to create identical clones either by splitting the early embryo or by nucleus transplantation. In the latter case the desired full complement of genes is inserted into a fertilized, denucleated egg.

These and other applications of cell mechanics and genetic technology contributed to the commercialization of a new class of environmental products. The following section provides a framework for classifying these research programmes in biotechnology as a prelude to an examination of early public policy responses.

ACTION OF GENES IN DEVELOPMENT

In view of what has just been set forth, how are we to conceive the operation of the genes in producing the characteristics? How do they operate in producing an adult from the fertilized egg? And how do different sets of genes produce individuals having different characteristics? On the intimate details of the operation of the genes little is known. Some of the main features in their method of action were taken up earlier in dealing with the relation of chromosomes to sex, though only the action of large groups, forming chromosomes, was there dealt with. From the relations brought out above, and in preceding chapters, taken in connection with the facts of embryological development, it is possible to form a conception of some of the main features in the action of genes in building up the body. These are summarized in the following.

The genes form a set of many diverse materials, differing from each other chemically and physically and having different chemical and physiological activities. They exist at the beginning of the individual's development in the nucleus of the fertilized egg, while surrounding them is a mass of cytoplasm. The genes interact with the cytoplasm, changing it, producing new materials that are incorporated into the cytoplasm. These varied cytoplasmic products of the different genes interact with each other, producing again new types of material. Some of the gross features of the interaction of genes and cytoplasm were described earlier. As there shown, the chromosomes (groups of genes) visibly take up material from the cytoplasm, alter it, and give it off again into the cytoplasm.

Though the genes thus interact with the cytoplasm, modifying it, they are not themselves used up in this process so as to lose their essential character. They supply material which combines with cytoplasmic material. But there is left an unchanged part of each gene, and this by division produces additional genes of the same type and action. As the cells divide, each chromosome divides, and each of its contained genes divides. One of the two halves of each gene goes after division to each of two cells produced. Thus each new cell produced contains the full set of genes, all those that were present in the fertilized egg. In most organisms it appears probable that this continues throughout at least all the early stages of development, if not to the end of development, so that all the cells contain all of the genes. For in the young embryo, after many hundred cells have been produced, it can be shown that any cell can produce many different parts of the organism. And in many organisms even in late stages there are cells that can produce any part, or the whole organism, as is shown by the facts of regeneration. Such cells must contain the entire set of genes.

In very late stages of development it is possible that in highly differentiated tissue cells some or all of the genes have become lost or modified, though there is no clear proof of this. Further, there are a few organisms of which it is to be observed that the cells that are to produce the differentiated body of the animal lose parts of their chromosomes, and so presumably some of the genes. The chromosomes in these cells are seen to break up, and portions of them are dissolved and absorbed by the cytoplasm while in the germ cells the chromosomes remain entire. This is known to occur in threadworms (Nematodes) and in certain insects. The relation of these processes to the way bodily development occurs is quite unknown.

But on the whole the evidence is strong that in most organisms the production of different tissues and organs by different cells is not due to their containing different genes, for cells that certainly contain all the genes produce many different parts in many different cases. The genes act differently in the different cells, depending on the relation of the different cells to each other, so that the different cells, though containing the same genes, produce different tissues and organs.

The diversity of the different parts of the body seemingly arises mainly, if not entirely, through the fact that different cytoplasmic products of the genes are separated into the different cells. Thus the cells become different in their cytoplasm. And this different cytoplasmic constitution of the different cells interacts differently in each case with the full set of genes that the nucleus contains, so giving anew different products. By a continuation of such action the different tissues and organs are produced. In each, the cells contain the same sets of genes, but differ in their cytoplasm. The differentiations of the body therefore lie mainly in the cytoplasm, though the diversities in the cytoplasm are brought into existence through the action of the genes.

Yet it is true that differences may arise within the body of a single individual through a change in the genes of some of the cells. As before seen if in a female Drosophila one of the cells loses one of the X-chromosomes, that part of the body derived from this cell shows the characteristics of the male, while the rest is female.

If the X that is lost carries dominant genes, while the remaining X has the corresponding recessive genes, then, in that part of the body whose cells have lost one X-chromosome, the recessive characters are manifested, while the remainder of the body is dominant. In this way are produced mosaic organisms, showing diverse hereditary characteristics in different parts of the body. Such mosaic diversities, however, appear to be of a very different type from the differentiation of the body into tissues and organs.Whether any differentiations in ordinary development are thus produced through loss or modifications of certain genes must be considered as yet uncertain, though it is entirely clear that most of them are not so produced.

But diversities between individuals, as we have seen, commonly result from differences in the genes that they contain. If two fertilized eggs begin with different sets of genes, the processes of changing the cytoplasm, of differentiating it into tissues and organs, differ in the two cases, so that individuals with diverse characteristics are produced. Very marked differences in characteristics may be produced even when the two sets of genes differ in but a single gene or pair of genes. In Drosophila, as before seen, if two eggs differ in a single pair of genes in chromosome IV, one of them produces individuals without eyes, while the other has normal eyes. If two individuals differ in an entire chromosome (X), one becomes a male, the other a female.

Such differences are due to the diverse chemical and physical action of the different genes. They produce differences among different cells, in the way set forth in earlier paragraphs. Also, they may produce diverse chemicals (hormones) which circulate through the body, thus causing changes in the development of parts with which they come in contact.

Different chemical and physiological effects, different development, may be produced by difference in balance among the genes. Two cells may contain the same kinds of materials (genes) throughout, but one of these contains certain materials in two doses (a pair of genes), while the other contains but one dose (an unpaired gene). In consequence the two cells differ greatly in development. In this way difference in sex is produced, with its manifold attendant differences in structures and functions. In many cases two genes of a certain kind produce a different effect from a single one, though in other cases they do not. That is, two centres of a certain kind of action may give different results from a single centre. This principle of the effects of difference in balance is presumably of great importance, though as yet not a very great deal is known as to its operation.

It appears possible that the operation of some or all of the genes begins in early stages of development and continues throughout. It is known that some genes produce their distinctive effects at the beginning of development, in the egg itself. Thus in the silkworm a difference of one gene is known to make a difference in the programmes of the egg. Two races differ in egg programmes, one having slate-coloured eggs, the other brown eggs. Crossing the two races shows that the difference in programmes is due to a difference in a single gene, for the descendants show 'unit difference' inheritance.

If, by crossing, the special gene from the slate-coloured race is brought into the eggs of the brown race, in the next generation the eggs are slate-coloured instead of brown, so that the 'slate' gene is active in the formation of the egg. Other genes are known to produce their distinctive effects somewhat later, but still rather early in development. Thus in mice, if one of the genes of a certain pair is modified in a certain way, this causes the hair of the individual to be yellow.

If both the genes of this pair are so modified as to produce the yellow programmes, the individual develops for but a short time, then dies while still an embryo. It is clear that these genes are active in the young embryo. Again, in plants, certain genes are required for the production of chlorophyll. If one of these becomes defective, chlorophyll is not produced. The young plant grows so long as the stored food in the seed is available, but, having no chlorophyll, it cannot make use of the sun's rays for the elaboration of nutrition; it therefore dies as a young seedling. Examples of the operation of genes during early development are abundant.

Some genes are required for laying the foundation of certain parts or organs; if these genes are altered, no such organ is produced. Thus in the fruit-fly a certain gene in chromosome IV is necessary, in its normal condition, for producing eyes. If that gene becomes defective, the eyes are completely lacking, although all the other genes that cooperate to produce the eye may be present in normal condition. This particular gene is required for the very foundations of the eye. Similarly, in vertebrates a certain gene is required for the production of the pigment which produces the colours of hair and skin. Many different genes cooperate to produce this pigment; most of them modify the type of pigment produced. But if this particular one becomes defective in a certain way, no pigment is produced; the animal is an albino.

A somewhat similar situation appears to be at the basis of some cases of feeblemindedness in man. In such cases the brain is not properly constituted; it does not function well. Study of inheritance shows that in some cases, at least, feeblemindedness is due to a defect in a single gene; there is 'unit difference inheritance. This single gene plays a necessary part in producing an efficient brain; when the gene is defective the brain is fundamentally imperfect.

The special and distinctive functions of some genes are not known to come into operation until late stages of development. Such, for example, are those that produce in Drosophila the distinctive pigments of the eye. Their visible effect does not appear until the eye is formed; although of course this action may have begun early in development.

In many such cases in which the conspicuous effect of a gene appears at a relatively late stage in development there are less conspicuous effects that indicate that the gene may be in operation throughout development. Thus individuals with white eyes, or with rudimentary or vestigial wings, are less vigourous than those with normal eyes or wings; they are less resistant and have shorter lives, and this is true for most of the types having recessive characters. Though the conspicuous special effect of the changed gene is late in appearing, the effect on the general constitution indicates that there may be an effect throughout development.

That this is indeed the case is indicated strongly by certain other facts. Sometimes a gene at a certain locus in the chromosome becomes entirely inoperative; it has either dropped out of the chromosome, or it has totally lost its action. Such cases are commonly known as deficiencies. Deficiencies are producible by the action of radiation on the chromosomes, so that large numbers of them have been studied.

A deficiency is detected in the following manner. A chromosome suspected of being deficient for a certain gene at a certain locus is brought into an individual in which the other member of the pair to which it belongs carries a recessive gene at that locus. If the chromosome is deficient in the suspected gene, the recessive gene is manifested, since there is present no dominant gene of that pair. Thus if an X-chro- mosome carrying white eye is brought into a female whose other X is deficient for the gene at the white eye locus (the locus 1.5), the female has white eyes.

In the case of almost any gene, if two chromosomes that are deficient for the same single gene are brought into the same fertilized egg, so that the egg lacks that gene entirely, the egg does not fully develop. In most cases it is not known at just what stage development ceases. But the fact shows that, in addition to its special effect on a particular characteristic, the gene is necessary if development is to continue at all beyond a certain stage. By ingenious methods Demerec has recently proved that if only a few of the cells of a developing embryo are deficient for a certain gene, those few cells fail to develop, though the rest of the body 'does develop.

This was demonstrated for deficiencies of many different genes in the X-chro- mosome of Drosophila. In only one of the many cases tested was deficiency for a certain part found not to be lethal, though even in this case it caused weakness. In the case of the white eye gene at I, 1.5, and in many other loci, deficiency prevented development. Altogether, it is clear that genes are active throughout development, some beginning early, others seemingly later, and that almost all genes are necessary in order that full development shall occur.

Changing one or more genes may cause from the beginning a change in the chemistry and physiology of the organism, altering its course of development and so giving rise to changed characteristics.

What kinds of Characteristics are affected by Genes?

The question as to what kinds of characteristics are affected by genes has been much discussed. Some have supposed that only a few characteristics depend on genes: only certain superficial matters, such as colours, and the forms of external features. Particularly with relation to man has there been dispute on this matter.

But on this question, positive knowledge is available, both for man and for other organisms. It is possible to answer fully the question: What kinds of characteristics are affected by genes?

The question can mean only: What kinds of characteristics can be altered by changing the genes? What kinds of differences in characteristics result from differences in genes?

There are several ways of answering these questions. The most direct is by the crossing of individuals having different genes. This results in substituting one type of gene for another, with consequent differences in characteristics. If the diversities in characteristics show Mendelian or sex-linked inheritance, this demonstrates, as we have seen, that they are due to differences in genes.

By this test, in many organisms (for example Drosophila), the following kinds of characters have been found to be affected by genes:

- Structures and forms of all parts (head, eyes, thorax, abdomen, wings, legs, internal organs).
- Colours and patterns of all parts. These depend on chemical processes, hence:
- Chemical processes of many kinds are affected by genes.
- Physiological conditions: health, vigour, life and death.
- Power of development: certain lethal genes stop development at a certain stage. A defective gene in the fourth chromosome of Drosophila prevents the development of the eye. Certain genes cause the body to grow to a larger size than usual. Many examples of the effects of particular genes on development have been given in earlier pages.
- Behaviour. Reactions to light in Drosophila are changed by alteration of a certain gene ('tan').
- Sex. Changing the genes changes the sex; this involves changes in all types of characteristics: in structure, pattern, programmes, physiology, behaviour.

Thus, in sum, it is clear that in animals, characteristics of all kinds, without exception, are affected by genes. There are no characteristics that are exempt from influence by the genes. Any characteristic can be altered by suitable alteration of genes. This is not surprising when it is considered that the genes are among the materials from which organisms are made, and that altering them alters the chemistry of the developmental processes. What is the situation in Man? What characteristics are affected by genes in Man? In man we have to deal with certain important types of characteristics that are hardly to be studied in other organisms. These are what we call mental characteristics: emotions, temperament, skill, character, intelligence, reasoning power, and the like. There are great differences among human beings in their mental characteristics, and in their capabilities. Are these things affected by genes? Often it is argued that they are not. Some have held that practically all differences in mind and character are due to environment, to education, influence of associates, tradition, state of civilization, and the like. On this matter there is positive knowledge which completely removes the matter from doubt. There are two methods by which it can be determined in man what kinds of characteristics depend on genes.

1. The first method is by observing what characteristics show Mendelian or sex-linked inheritance. This method is not entirely satisfactory in man, since;
   a. It is not possible to carry on experimental breeding in man,
   b. This method can hardly be applied to the important mental characteristics. Yet by this method many characteristics of man have been shown to depend on genes. Among the characteristics showing sex-linked or Mendelian inheritance, and so depending on genes, are the following in man:
      i. Programmes of hair and eyes; form of many facial features; structure of the hands and feet; form of the body, whether stout or slender; chemical constitution of the blood; efficiency of the brain (whether the individual is normal or feebleminded); condition of the eyes and sense of sight (colourblindness, short-sight, night-blindness and other conditions); sense of hearing (deafness, etc.); sex, and many other characteristics.
2. A second method gives much fuller and more detailed knowledge of what characters are affected by genes in man. This is by the study of identical twins, as compared with other individuals. This is a most important method, and it will therefore be discussed in some detail.
   It depends on the folio wing facts:
   a. In man, most individuals are derived from separate germ cells, and therefore have diverse sets of genes. For in forming germ cells, as before described, reduction of chromosomes and genes occurs, so that practically every germ cell differs in its gene combination from every other. When by fertilization two different germ cells unite, the diversity of gene combination is increased still more. The result is that no two individuals formed in this way have the same combination of genes.

b. But rarely there occur cases in which a single fertilized egg produces two individuals (twins). Such individuals are formed by division of all the genes, so that the two twins contain the same gene combination. These are the twins known as monozygotic or identical twins. In another type of twins, the two are derived from separate eggs; such dizygotic twins have diverse gene combinations.

By comparing the similarities and differences of identical twins with those in other individuals, and particularly with those of twins of the other type, it is possible to find out what characteristics are affected;

- By having all the genes alike,
- By having different combinations of genes.

'Identical twins', or offspring all having the same genes, occur in many organisms. In lower animals (Protozoa, plants, Hydra, worms, etc.), a single individual often divides into two or more, by Vegetative* reproduction. All these individuals have the same sets of genes, since in vegetative reproduction every chromosome and every gene divides. Such individuals are very exactly alike. They have been studied a great deal. In such a Protozoan as Paramecium, a great number of individuals are produced in this way, all closely alike. But when conjugation takes place, this gives new gene combinations, and the individuals after they have conjugated are very different.

In one of the vertebrates, the armadillo, identical twins are formed in essentially the same way that they are in man, but are formed regularly. A single fertilized egg, after developing some distance, divides into four, so that four 'identical twins', or quadruplets, are produced at each birth. The four are extraordinarily alike even in fine details. But twins belonging to different sets (and so having different gene combinations) are often very different.

IDENTICAL TWINS IN MAN

Identical twins are produced in man by the division of a single egg into two, so that they may be called monozygotic or 'one-egg' twins. There occur also dizygotic or 'two-egg' twins, produced from two separately fertilized eggs; these are often called fraternal twins. Methods have been discovered by which it is possible to determine to which class any pair of twins belongs.

Identical twins in man are of enormous interest; they supply exactly what is required for discovering what characteristics depend on genes. They are two individuals that are made throughout from the same set of materials, from the same set of genes. Every single gene of one has its exact duplicate in the other. The two therefore have the same heredity; the differences between them are due to environment alone. They show how like or how diverse will be two human beings that do not differ in their genes, in their heredity, although they may differ in the conditions of their lives, in the experiences they undergo. Identical twins thus form one of the most interesting and illuminating of the phenomena that nature presents to us. They yield the key to that ancient question, the relative role of nature and nurture: of heredity and environment. The existence of the two-egg twins, with their diverse sets of genes, for comparison with the one-egg twins, having identical sets of genes, completes what is required for discovering just what resemblances and differences in human beings are due to identity and diversity of genes. The conditions to which the two members of a pair of twins are subjected are as much alike in the two-egg twins as in the one-egg twins. Therefore we can be certain that any differences between the two types of twins are due to differences in the genes, not to diversities of environment.

Comparison of one-egg twins (having the same genes), with two-egg twins (having different genes): Identical twins are extraordinarily alike in all physical respects. They are always of the same sex. Their faces and Figs are so much alike that it is difficult to distinguish one from the other. They are alike in eye programmes, hair programmes, skin programmes. Their facial features are of the same form and relative size. Their teeth are alike, even to the irregularities. Their stature and weight are very similar. Their finger prints are as much alike as those of the two hands of the same individual.

In all these respects the two-egg twins are often very different. They are no more alike than any other two members of the same family.

Identical twins are also alike in weaknesses, in tendencies to certain diseases. If one has tuberculosis or other constitutional disease at a certain time of life, the other one usually has it also at about the same time, even though the twins live apart. But two-egg twins do not show these resemblances.

All this demonstrates that all these physical characteristics in man depend on genes. They are alike in two individuals whose genes are alike, diverse in two individuals whose genes are diverse.

One-egg twins at times show certain physical differences, resulting either from certain features of the process of splitting the original egg from which the two were produced, or from the fact that one of the twins has been injured by accident or disease. From the former cause arise the facts that:

- Sometimes one of the twins is right-handed, the other left-handed,
- One is sometimes more vigourous than the other. If the splitting of the original single egg occurs very early, before the right and left sides have become diverse, the two twins are alike in 'handedness' and in vigour; if the splitting occurs later they may be diverse in these respects. The effects of diversity of environment on twins we take up in a later chapter.

MENTAL CHARACTERISTICS IN TWINS

In a number of cases the two twins from a single egg have been separated while very young before the age of one or two years, and have lived apart afterwards. In such cases the genes of the two individuals are alike, while their environments have been different. What is the situation as to their mental characteristics?

Several cases of this sort have been very carefully studied by Muller and by Newman. We shall have occasion to look at the results of this study in some detail later. Here we may summarize briefly from our present point of view what was discovered. One-egg twins that have lived apart show the same close resemblance in physical characterization that is found in such twins as have lived together. They show also resemblance in mental characteristics: in emotions, temperament, and in 'intelligence' as measured by psychological tests. In certain cases, there are found some dissimilarities in mental and emotional characteristics, depending on differences in education and experience in the separated twins.

Thus it is clear that:

- The genes have great influence on mental characteristics, emotions, temperament, and the like, but that
- These things may be modified by different education and experience.

An illuminating study of the mental and moral characteristics of twins was made by Johannes Lange. He found in the prisons of Germany, or in the prison records, thirty persons, each of whom was a member of a pair of twins, the two members in each case being of the same sex. He traced and examined the other twin of each pair. Thus one twin was chosen because he was a criminal; the other twin was taken as he chanced to be. As it turned out, thirteen of the pairs were single-egg twins, while the other seventeen were two-egg twins.

Of the thirteen pairs of one-egg twins, one of which was known in each case to be a criminal, it turned out, upon investigation, that the other also had a criminal record in 10 out of the 13 cases. It is clear that if one such twin comes in conflict with the law, the other one, identical in original nature, has small chance of escaping that fate. But with the two-egg twins, made up of diverse original materials, the situation was different. Their environments were as similar as were those of the one-egg twins. But of the 17 pairs of this kind, in only three cases was the other member of the pair found to have a criminal record. In the other 14 pairs the fate of the two was quite diverse. One was a criminal in each case; the other was not.

Thus it is clear that everything that helps to determine whether a man shall be a criminal or a good citizen is deeply affected by the materials of which the individual is made, by his 'heredity'. Mentality, morality, conduct, career, fate, are affected by the hereditary materials with which the individual starts life.

The truth of this conclusion came out still more strongly when the detailed records of the twins were studied. In the case of the single-egg twins, the two members conducted themselves as duplicate personalities might be expected to do. One such twin was a burglar; on looking up his brother's record, he was found to be a burglar too. In another pair, one twin was a high-powered financial swindler, fraudulently collecting large sums from his dupes. And so also was the other twin of that pair.

Besides the pair that were burglars, and the pair that were swindlers, another pair of one-egg twins both 'committed puerile offences against the property laws'. Another consisted, in the words of Lange, of 'guttersnipes, but good fellows at heart', though they 'cannot stand alcohol; it develops frenzy and draws the knives from their pockets'. Another pair described 'have too little sense and will power', while in still another, both 'are lacking in all human feeling except for their own unpleasant selves'. And so goes the account throughout the entire list; the similarity in conduct is astonishing. Lange says, 'In all these pairs the type of crime is identical, the criminal careers begin at about the same age, and the behaviour of both members in court and in prison corresponds absolutely.'

But in the two-egg twins, that started life with different parental materials, with different heredity, the situation was different. As we have seen, 14 out of the 17 had totally different careers, one being a criminal, the other not. And there was no such detailed parallelism in the records of the two as was found in the one-egg twins.

As before remarked, the similarity of environment was as great in the two-egg twins as it was in those that came from one egg. But this similarity of environment did not result in identity of fate or in detailed parallelism in careers. It was the identity of original nature, of the materials out of which they are made, that gave the same careers and the same fate to the one-egg twins. They were almost as much alike in their behaviour, their mentality, their career and their fate, as they were in their physical features in which, like other one-egg twins, they were almost indistinguishable.

It is certain, therefore, that all the things that affect character and conduct are deeply influenced by the hereditary materials. There is no characteristic or quality of human beings that is exempted from its influence. This conclusion is confirmed by all the many studies that have been made on the two types of twins. And it is in harmony too with all that we know of the science of genetics. Experimental work on the breeding of other organisms shows, as we have seen, that they have no characteristics of any kind that are not affected by the genes that they receive at the beginning of their lives.

Are characteristics influenced also by other conditions, by the environment? The fact that all are affected by genes does not at all imply that they may not be affected in other ways.

SPECIAL ACTION OF A GENE

- Particular genes affect mainly, though not exclusively, some particular part of the body, or some particular function.
  Thus, in Drosophila there is in the X-chromosome, at the locus I, 1.5, a certain gene which in its normal condition takes part in producing the red eye programmes. This gene has become modified in different cases in a considerable number of different ways, and any modification alters the programmes of the eye. The different

modifications give the many different alleles listed earlier among the eye-programmes characters of linkage Group I. Other genes have their chief effect on the body programmes or on the wings or on some other part. Particular genes in man affect the blood type of the individual; others affect the eye, the body form, and so on. Each gene as a rule has its main function in connection with the development of some particular part.

- Yet a given gene often produces effects on several parts or functions. For modification of a single gene may cause a change in the characteristics of a number of different parts. Some examples of this may be given. The character bent wings, belonging to linkage Group IV of Drosophila, is due to the modification of a single gene in chromosome IV. It causes the wings to be bent dorsally at their base, and also it causes the legs to be shorter than normal, owing to the abnormal shortness and thickness of the meta- tarsal joint.

  In animals showing the character vestigial wings (due to modification of a gene in chromosome II), the following peculiarities appear: the wings are degenerate, only their bases being present, and these are held at right angles to the body. The balancers are likewise degenerate. Two rear bristles on the scutellum are farther apart than usual, and they stand up straight instead of inclining backward. The flies hatch about two days later than in the normal wild flies. They are not quite so vigourous as the normals.

  As noted in the preceding chapter in connection with linkage Group I, modification of a certain gene in the X-chromo- some causes;

  a. The body programmes to be tan instead of grey,
  b. Causes the individuals to lose the usual positive reaction to light.

  In any detailed account of the effects of particular genes, it will usually be found that each gene has thus several different effects, often on diverse parts of the body. The effect on which the name given to the gene is based is commonly merely that effect that is most conspicuous to the eye.

General or Constitutional Effect of a Gene

- In addition to the special effects on certain structures or functions, modification of any gene has also a general effect on the bodily constitution. In Drosophila the 'wild-type' flies, with unmodified genes, are vigourous and long-lived. Practically all modifications of the normal genes make the animals less vigourous, so that individuals showing any of the characteristics listed in the preceding chapter are weaker than the normal or 'wild- type* individuals, and have a higher rate of mortality. Thus practically all the modifications of normal genes are defects, a fact to which we return in a later chapter.

  In view of the facts set forth in preceding paragraphs it has been said with much probability that 'every gene affects the development of the entire body.

- Thus apparently each gene supplies material which in the process of development enters into reaction with materials from all the other genes; it thus modifies many or all parts of the body, but affects certain parts more conspicuously than others.
- As before set forth, a particular single gene may become modified in many different ways, in different individuals or different chromosomes, each modification causing a changed characteristic. The classical example of this is the series of different eye colours (multiple alleles) found in different individuals of Drosophila, each programmes being due to a different modification of the gene located at the point I, 1.5 in the X-chromosome. This gene in its normal condition produces (by interaction with other eye-programmes genes) the normal red programmes. The various modifications of the gene give rise to the colours before mentioned: coral, blood, eosin, cherry, apricot, buff, tinged, ivory, white. Many such series of multiple alleles are known in the fruit- fly and in other organisms.

  In the fruit-fly the following illustrate their nature:

  - *Scute*: The gene scute is close to the upper (or left) end of the X-chromosome. It exists in 20 to 30 different modifications, each of which causes a slight change in the distribution of the bristles on the body.
  - *Cut*: A considerable number of different alleles are known of the cut gene, located at I, 20.0. All of them cause changes in the outline of the wings, as if bits of the wing were cut off.
  - *Vestigial*: The gene vestigial at II, 67, which reduces the size and changes the form of the wing, has two other alleles, antlered and strap> giving wings of other forms and sizes.
  - *Garnet*: At the locus I, 44.4 are several alleles which change the programmes of the eye in diverse ways. Some of these are known as garnet, salmon, garnet 2, garnet 3.
  - *Pink*: The eye colours, pink, peach, pink 3, pink 4, pink 5, are a series of alleles located at III, 48.0. Series of alleles are known also in other organisms. Thus in the rodents (rats, guineapigs, rabbits), series of different alleles are known, yielding different colours of the hairy coat. In maize there are several allelic series modifying the programmes of parts of the plant. Presumably any gene may, by successive modifications, produce a series of alleles. As mentioned in an earlier chapter, all the members of an allelic series affect the same character or part of the body, yielding different modifications of that character.

    Further, they show certain peculiarities of inheritance as compared with different genes that are not alleles. Since they are all modifications of the same gene, only one of the alleles can be present in a single chromosome, and for this reason only one of them can be carried by a single gamete (except in cases of non-disjunction). Further, only two members of a given allelic series can be present in any diploid individual or zygote.

There are thus two types of diversities among genes,

a. On the one hand, there are the different genes occurring at different loci in the chromosomes, the diverse genes, shown on the maps affecting many diverse characteristics.
b. On the other hand, there are these different modifications of a single one of these genes the multiple alleles.

- Any given single character, inherited in the sex-linked or the autosomal manner, is a product of the interaction of many genes, located in different parts of the genetic system.

  As an example of this, the programmes of the eye in Drosophila may be taken. As shown in the preceding chapter, the programmes of the eye depends on many genes scattered through the different chromosomes. The number of different genes directly affecting the programmes of the eye is indeed still greater than this, and in addition there are many other genes that affect eye programmes indirectly, through the fact that they affect the structure of the eye: for example, the gene 'eyeless* in the fourth chromosome pair. If this gene is modified in a certain way, no eyes are developed, consequently no programmes.

  Similarly, many different genes affect the form and structure of the wings, the venation of the wings, the structure of the eye, the programmes of the body, the form of the body, the structure and distribution of the bristles. Indeed, this is presumably the situation with relation to any characteristic whatever; every characteristic is influenced by many genes.

- Altering any one of the numerous genes that cooperate to produce a given characteristic alters that characteristic. Any characteristic may, therefore, be changed by altering any one or more of many different genes.

  Thus, consider what must be the situation among the genes to yield the normal red eye programmes in Drosophila. To produce this result all of the 19 (and more) genes that affect eye programmes must be present in the usual or normal condition. If any one of them is changed (the rest remaining unchanged), the programmes of the eye is altered.

  Changing in a certain way a gene at I, 1.5 all other genes remaining normal, the eye programmes is changed to white. If we change the gene at I, 33 (the rest remaining normal), the programmes becomes vermilion. If we change the gene at II, 54.5 (the rest remaining normal), the programmes of the eye becomes purple. Thus one could go over the entire 19 genes; changing one after the other; each one changed alters the eye programmes.

Again, consider what must be the situation in the genetic system to yield one of the recessive colours: for example, vermilion eye programmes. To produce this programmes, 18 of the genes must be in their usual normal condition, while a single one, at the point I, 33, must be altered. If, in addition to this one, another is altered (for example that which by itself yields purple, at II, 54.5), a different programmes is produced. Altering one gene changes the programmes in a certain way; altering two or more changes the programmes in a still different way. Similar relations would be found with respect to most or all characteristics.

- *Multiple Factors*: The different genes on which a given character depends may have qualitatively different effects; such is the case with the different genes that affect the eye programmes in the fruit-fly. In other cases the different genes affect the characteristic only quantitatively. Thus dimensions often depend on many different genes. Modifying one of these or substituting another for it changes the dimension; modifying an additional one changes it still more, and so on. In such cases the character is commonly said to depend on 'multiple factors'.
- *'Duplicate Factors*: In some cases a character may depend on the presence of two (or more) genes of a particular type in such a way that both of these genes must be present in order that the character shall be produced. The character is then said to depend on duplicate genes or factors. In the plant known as the shepherd's purse, the common type of pod is triangular, but some plants have an elongated oval pod. This oval pod occurs only when two particular recessive genes are present in the homozygous condition. Thus if the two dominant genes which produce the triangular pod are A and B, while the two recessives that give the oval pod are a and b, the oval pod occurs only in the plants having the genes aabb, while all the other combinations give the triangular pod. Similar cases of duplicate genes are known in a number of organisms.
- 'Modifying Factors'
    a. In some cases the effect of some of the genes is merely to modify in detail a characteristic that is due primarily to another gene. Such are known as modifying genes or factors. Thus, in Drosophila, a certain modified gene at I, 1.5 causes the eye to assume the pinkish programmes known as eosin, but the depth of this eosin programmes depends upon certain other genes present in other parts of the genetic system. One of these, known as cream b, is located at II, 22.5; another, cream III, at III, 36.5;

and several other such genes modifying the eosin programmes are known. If these genes are present in the normal or unmodified condition they do not affect the eosin programmes; when modified in certain ways they do. Again, in the rat, the presence of a certain gene causes the fore part of the body to be coloured brown, forming what is called a hood) the rest of the body being white. The extent of this hood depends on the presence in the individual of a number of other genes; some of them increase its extent, others decrease it. Such modifying genes or 'modifying factors' play a role in many characteristics. They may produce no visible effect except in the presence of a particular character that they modify. Thus the modifiers of eosin eye programmes produce no visible effect in an individual with red eyes.

b. The same phaenotypic character that is, the same characteristic in outward appearance may be due in different cases to the alteration of diverse genes, in different parts of the genetic system. This is illustrated in numerous cases in Drosophila.

Thus, modification of a certain gene at II, 54.5 causes the programmes of the eye to change from the usual red to a darkish purple; the modified gene giving this programmes is known as purple. But in other cases what is outwardly the same programmes is found to be produced by change in other genes. One of these is at I, 44.4; to distinguish it from the one just mentioned it is called garnet. A third gene that produces practically the same eye programmes is found at III, 49.7; it is called maroon. Similarly the eye programmes 'salmon', resulting from a modified gene at I (again at 44.4), is indistinguishable from the eye programmes 'rose', resulting from a gene modification at III, 48.0.

Other cases of this sort could be given. In such cases the same phaenotypic character will be found to be inherited in different ways in different cases, depending on the location of the modified gene to which the character is due. If the gene is in X, the inheritance will show itself to be of the sex-linked type; if in II, it will be autosomal and of the linkage Group II; if in III, it will be autosomal and of the linkage Group III.

CONSTITUTION OF THE CHROMOSOMES THE GROUPS OF CHARACTERS LINKAGE

Each of the chromosomes (save in some cases the Y-chromosome) is composed, as we have seen, of many diverse genes. In the wild, unmodified individuals, in the fruit-fly, the genes are such as produce the 'normal' or 'wild-type' characteristics of the individuals. The genes of any chromosome become modified in the course of time, in the process known as mutation, which will be dealt with in a later chapter. The modified genes produce altered characteristics, most of which are recessive, while the normal characteristics, when mated to the modified characteristics, are dominant. A few of the modified characteristics, however, are dominant when mated with the normal or wildtype characters. The modified characteristics resulting from mutation are in most, if not all, cases defects, as compared with the normal characters. Thus the normal, dominant characters form a standard condition, of which the defective and recessive mutated characters are modifications.

The diverse genes of any chromosome are designated, so far as they have received names, by the name of one of the modified characters which it helps to produce. Thus the gene yellow, near the upper or left end of the X-chromosome of Drosophila, produces in its mutated condition the modified body programmes yellow; while in its unmodified condition it produces the normal grey programmes of the wild fly. In the maps of all the names apply to the modified or mutated condition of the gene. For every such modified gene there is a corresponding unmodified normal or wild-type gene. It is the custom to designate the wild-type gene from which any modified gene is derived by writing with the name or abbreviation for the modified gene the sign +. Thus yellow + or y + or + y signifies the normal wild-type gene that produces grey body programmes; white + or w + signifies the normal gene located at I, i.5, which produces red eyes in place of white ones.

The genes of any one of the chromosomes, as X, or a given autosome, become modified in many different ways, so that different examples of the same chromosome have different sets of genes. The diversity of genes in the chromosomes of different germ cells causes diversities in development, with resulting different characteristics in the individuals produced. The characteristics produced by the action of the genes of any chromosome follow in the individuals of successive generations the distribution to the offspring of the genes that produce them. Thus many characteristics follow the course of the X-chromosome; many follow each autosome; a few may follow the Y-chromosome.

Since a large group of characteristics follows each chromosome, the result is that characteristics pass from parent to offspring in groups. As an X-chromosome passes from the father into his daughters, all the genes, and all the characteristics that they produce, pass together into the daughters. Here, of course, some of the recessive characteristics may remain hidden in the presence of dominant genes in the second X- chromosome carried by the daughters. But all may reappear anew when in a later generation this X-chromosome is again by itself in a male. Except in so far as there is exchange or crossing-over with the other chromosome of the pair, all the characteristics that depend on the genes of a particular chromosome pass together into the individual that receives that chromosome, and this continues for generation after generation.

Thus, in Drosophila, suppose that a male has white eyes and yellow body, two characteristics connected with genes in the single X-chromosome. He is mated with a female that has red eyes and grey body, again owing to genes in the X-chromosome. In the daughters of these two, the two types of X-chromosomes are present together. These daughters produce offspring in a third generation. Some of their sons receive the X-chromosome that came from the original male; these will have white eyes and yellow body. Other sons receive the X-chromosome that came from the original female; these have red eyes and grey body. Only a very minute proportion of the offspring have the reverse combinations of characteristics white eyes with grey body, and red eyes with yellow body.

The two characters originally together white and yellow appear to be linked, and the same is true of the other combination, red and grey. The original combinations do not readily separate.This linkage of characters in heredity was observed before it was known that characteristics depend on chromosomes. Many theories were invented to account for it, until it was discovered that it is due to the fact that the linked characters depend on genes that are in the same chromosome. All characteristics that depend on the same chromosome form what is called a linkage group. Since there are different numbers of chromosomes in different organisms, there are correspondingly different numbers also of linkage groups.

These linkage groups are of great importance in heredity. They reveal to us many things about the nature and action of the genetic system. We shall therefore examine these groups in a typical organism, selecting for this purpose the animal in which they are best known, namely, Drosophila melanogaster.

THE GROUPS OF LINKED CHARACTERS IN DROSOPHILA

The fruit-fly has, as we have seen, four pairs of chromosomes. On the genes in these four pairs of chromossomes depend several hundred known inherited characteristics. These many characteristics form four linked groups, according to their dependence on genes in the X-chromo- some, or in the chromosomes II, III or IV. It is to be noted that all of the characteristics that depend on genes in either one of a given pair of chromosomes constitute a single linkage group, since all those characteristics may become connected with the genes of a single one of the chromosomes. Thus the number of linkage groups is the same as the number of pairs of chromosomes.

In the course of breeding successive generations, linkage of charact-eristics shows itself in the fact that certain characteristics (dependent on genes in the same chromosome), which are present together in one parent, are present together also in the individual descendants. This is best seen on comparing grandchildren with grandparents. In this case one grandparent had the combination white-yellow, the other the combination red-grey. The same two combinations reappear in the grandchildren (seen particularly in the grandsons). In addition there are a very few 'cross-overs', having the new combinations white-grey in some individuals, red-yellow in others. But almost all the grandchildren get both the characteristics from the same gandparent.

If the original grandparents have the other combinations, grey-white and yellow-red, then the majority of the grandchildren show these combinations.

Thus just what combination of characteristics the individual grandchildren have depends on what characteristics were united in the grandparents. But this is the case of course only for characteristics whose genes lie in the same chromosome. If two characteristics depend on genes that are in different pairs of chromosomes, then the combinations that were present in the grandparents are no more frequent in the descendants than are the reverse combinations.

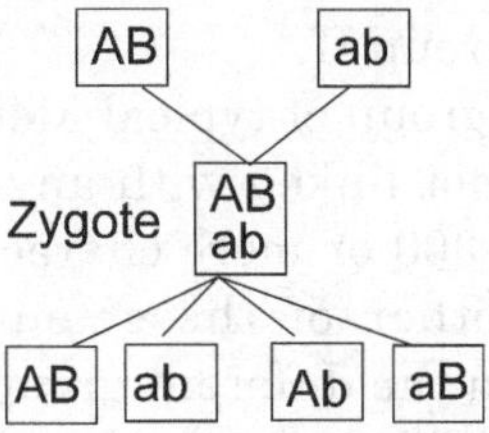

Fig. Diagram of the Results of Linkage, with Relation to the Gametes.

It is in many respects most convenient to think of linkage in terms of the gametes. The combination of genes that are present in the gametes can often be determined by examination of the offspring produced. With relation to the gametes, linkage can be characterized as follows. When an individual (zygote) is formed by the union of two gametes, each bearing in one of its chromosomes a certain combination of linked genes, this same individual later produces gametes of which The two gametes above carry respectively the linked genes AB and ab.

The zygote produced by their union carries both AB (in one chromosome) and ab (in the other). Then the majority of the gametes formed by this zygote carry again the linked genes AB or ab; only a minority cany Ab or aB. the majority have these same combinations of linked genes. Often also there is a minority of gametes that have the genes combined in a new way, as a result of crossing-over.

In the fruit-fly, crossing-over occurs only in the females, not in the males. Thus if a male individual is formed by the union of two gametes carrying certain combinations of linked genes, then later such a male produces only gametes having the same combinations of linked genes. This fact is extremely convenient for determining what characteristics are linked in Drosophila.There are many other organisms in which crossingover is not limited to one sex. By breeding for successive generations and by the study of exchanges between the chromosomes, the following groups of linked characters have been found in the fruit-fly.

- Group IA large group of sex-linked characters, more than 100 in number.These all follow in heredity the distribution of X-chromosomes, as described in earlier pages. Any two or more of these characters may be linked together and connected with the same X-chromosome.

- Group IIA large group of typical Mendelian characters, 50 to 100 or more in number. These follow the method of distribution of autosomes, and any two or more of the group may be connected with the same autosome.
- Group III.Another large group of typical Mendelian characters, 50 to 100 or more in number, following the method of distribution of autosomes. Any two or more of this group may be connected with the same autosome, but a character of Group III is not linked with any character of Group II.
- Group IV.A small group of typical Mendelian characters, only 5 or 6 known.These are not linked with any of the characters of Groups I, II or III.All of the 300 or more characters known in this species fall into one or the other of these four groups. The nature of the characters found in the different groups is considered on later pages.

Thus here the number of linked groups is the same as the number of pairs of chromosomes. There are three large groups and one very small one; likewise there are three large pairs of chromosomes and one very small pair. It is natural to conclude, therefore, that the groups of linked characters correspond to the pairs of chromosomes, and that the characteristics in each group are connected with some particular pair. As it turns out, there is conclusive evidence that this is true. The evidence for each of the groups may be summarized as follows:

All the characters in Group I are sex-linked; that is, they follow exactly the distribution of the X-chromosomes. Any character in this group follows a particular X-chromosome and its descendants, wherever that chromosome goes. The characteristics in Group I are thus certainly dependent on genes in the X-chromosomes.

The small group of characters known as Group IV has been proved to be connected with the small chromosomes of pair IV. This has been demonstrated in the following manner. By non-disjunction of this fourth pair of chromosomes, some individuals are produced having but one small chromosome (IV). That this is the situation can be seen under the microscope, in properly prepared material.

These individuals with but one of the small chromosomes have the dominant normal or 'wild' characteristics. Such individuals are mated with individuals containing one of the recessive characters of Group IV. One of these recessive characters is 'bent wings'; another is 'eyeless' the individual having no eyes.

Half of the germ cells from the individuals that have but one of the fourth chromosomes are without this chromosome, while half of them have it. These two kinds each unite with germ cells carrying the recessive character 'bent', or 'eyeless'.

There are thus produced individuals, half of which have but one of the fourth chromosomes, while the other half have two. In those which contain two, all the characteristics are of the dominant normal type; the recessive character bent or eyeless is not manifested. But in the individuals having but one fourth chromosome, the recessive character is manifested: the individual shows bent wings, or is eyeless. This demonstrates that the characteristics of Group IV are connected with the small chromosomes.

Thus the large linkage Groups II and III are left for the two large V-shaped chromosome pairs, II and III. These chromosomes can be broken into two or more pieces by subjecting them to radiations. When this is done, the characters of the corresponding linkage groups are no longer linked together.Thus in Group III the two genes scarlet and sooty are linked (completely in the male). But when the corresponding large chromosome (III) is broken into two pieces, scarlet and sooty are no longer linked; they may pass to different offspring. They are evidently connected with the two separated pieces of chromosome III. In the same way the other large chromosome (II) may be broken, and this causes the characteristics of the linkage Group II to separate into two groups. Minute study shows that the chromosomes II and III differ a little, in size and form, so that with practice it becomes possible to distinguish one from the other under the microscope.

Thus it is proved that each of the four linkage groups is connected with one of the four chromosome pairs; the four groups correspond to the four chromosomes. In a number of other organisms the number of linked groups of characters has been determined, and also the number of pairs of chromosomes. In every case in which this has been fully worked out, the number of linked groups is the same as the number of pairs of chromosomes. The numbers of groups and of pairs in certain organisms are as follows: Drosophila melanogaster 4, Drosophila virilis 6, garden pea 8, sweet pea 8, maize10. In man there are twenty-four pairs of chromosomes, including the X's. It is to be expected therefore that there will be found 24 groups of linked characters, one group sex-linked, the other 23 groups typical Mendelian, in their inheritance. But it will be a very long time before these 24 groups are identified.

GENETICALLY ENGINEERED LIFE-FORMS IN THE ENVIRONMENT

The idea of controlling and improving upon nature has been a persistent theme in the history of technology. It has always been met with its skeptics, who are no less zealous in their opposition to the hubris of science as the proponents of change are to their attack on nature worship. At one end of the continuum are those that argue there is nothing in nature that cannot be improved upon. At the other extreme are those who maintain that evolution, without human intervention, results in the optimality of life processes. Nature knows best, according to the popular aphorism.

When molecular genetics gave birth to gentechnics, there were already numerous examples of efforts to control and harness nature that had resulted in mixed outcomes. The mining and burning of coal and oil has resulted in new forms of pollution and has been a major contributor to the "greenhouse" effect or the warming of the earth.

Synthetic, non-biodegradable chemicals, which were responsible for the second American industrial revolution, have become ubiquitous residents in the landscape posing remediation problems of enormous magnitude and economic burden. The agricultural system has become hostage to chemical pesticides that continue to pose risks to human health and disrupt natural ecosystems.

The early promise of biotechnology included products and processes designed to "fix" the undesirable outcomes of other technologies.

It also incorporated research to improve upon existing plants and animals as food sources. In preparation for reviewing the expectations of biotechnology during the first decade, have classified the environmental applications into five categories: remediation of ecologically harmful technologies; improvement of existing plants; modification of animals; microorganisms to mine the earth; biological sensors.

REMEDIATION

Pollution Cleanup

Research efforts are directed at genetically engineering microorganisms that could degrade oil spills, detoxify chemical pollutants in soil and water, and improve biodegradation in wastewater treatment plants. The problems of off-shore oil spills, toxic wastes at Love Canal, New York, and Woburn, Massachusetts, polychlorinated biphenyl (PCB) contamination of the Hudson River, dioxin in Times Beach, Missouri, and scores of other cases highlighted the importance of dealing with toxic chemicals in the environment.

With the tools of classical microbiology, plasmid engineering, and recombinant DNA techniques at their disposal, scientists began experimenting with microorganisms that break down complex hydrocarbons. Early studies showed promise of a bacterium that could digest crude oil.

In 1980 General Electric won a patent claim for an oil-degrading strain of Pseudomonas, the first such claim issued for a microorganism *sui generis*. Nine years later, the B. F. Goodrich Co. received a patent for a novel plasmid that degrades ethylene dichloride, a compound found in many toxic waste dumps.

Other investigations sought microbes that could neutralize the synthetic herbicide 2,4,5,T (an ingredient in Agent Orange) and PCBs, as well as organisms that could degrade explosive substances and soil contaminated with nuclear materials. The disposal of radioactive wastes has been a persistent problem for the nuclear industry and the Department of Energy. Community resistance to the siting of low-level radioactive waste land fills remains high. The prospect of a biotechnological solution to radionuclides could soften the public's opposition to nuclear power.

Biological Alternatives to Agricultural Chemicals.

Promotional materials emanating from the new biotechnology industry spoke enthusiastically of replacing chemical pesticides with biological alternatives that are species-specific and safe for humans and other living things. Some of the new firms embraced the rhetoric of the environmental movement in describing their research agenda and their concern for things ecological.

Pesticides are necessary to grow crops economically and preserve the quality of our food supply. Worldwide, a staggering $14 billion are spent annually on crop protection products–$5 billion in the U.S. alone. Despite these huge expenditures, many destructive pests go unchecked and cause major crop losses.

The pesticide industry is currently dominated by manufacturers of synthetic chemicals. However, the interplay of environmental and economic forces is creating tremendous pressure for new products which are effective and economical while being safe to use and being compatible with the environment. Biological products are ideally suited to fill the market needs.

Among the major chemical companies, Monsanto was recognized for its strong development effort in agricultural biotechnology. Monsanto's R&D programme included the development of a tomato plant genetically engineered with a protein from the naturally occurring bacterium Bacillus thuringiensis that is supposed to resist insects.

Several environmental groups, including the Environmental Defence Fund, the National Audubon Society, and the National Wildlife Federation supported a project of the firm Crop Genetics that involved injecting a microbe called Cxc into the seedlings of corn. This microbe, which ordinarily lives in corn plants, was endowed with a gene from Bacillus thuringiensis, the protein product of which is toxic to certain corn pests.

Another group of agricultural research programmes was directed at slowing down the escalating use of artificial chemical fertilizers, which run off into surface waters and foul lakes and streams by the buildup of organic matter–a process known as eutrophication. The expectation was that plants or symbiont microrganisms could be genetically engineered to fix their own nitrogen from the vast reservoir available in the atmosphere.

Agracetus began marketing a microbial soybean seed treatment–a product with genetically modified strains of *Bradyrhizobium japonicum* to improve nitrogen-fixing ability. Another strategy involved transferring expressible nitrogen-fixation (nif) genes of organisms like *Klebsiella pneumoniae* into the chloroplasts of plants, endowing the latter with a selffertilizing capacity.

Agricultural Harvesting and Distribution.

Fruits and vegetables are frequently harvested in a preripened stage to accommodate harvesting machines and mass distribution networks. However, the quality of the produce is often affected by being cut off from critical enzymatic changes that take place during the natural ripening process: "While today's tomatoes are the most impressive since the dawn of produce in their resilience, disease-resistance, and travel-hardiness, they are also the most inedible."

Some genetic engineering research is directed at modifying crops so that they will be compatible with the harvesting technologies while also retaining the taste and quality of naturally ripened produce.

The challenge for biotechnology is to recapture the taste of produce as it was when grower and consumer were in close proximity, and to accomplish this without disturbing the existing agricultural system and by a means that does not compromise the environment.

IMPROVEMENT OF PLANTS

ADAPTABILITY TO HARSH ENVIRONMENTAL CONDITIONS.

Agricultural crops are adversely affected by many environmental factors, including frost, wind, mineral toxicity, soil acidity, salinity, and of course insect pests. A number of research programmes were devoted to redesigning plants to be more adaptable to harsh environmental conditions. Kenneth A. Barton and Winston J. Brill saw the possibility of mixing and matching in nature by transferring the stress resistant qualities of weeds to food crops. Among the harvests expected of the technology are disease and drought-resistant strains, self-fertilizing seeds, crops with tolerance to high salinity, and plants with enhanced photosynthetic qualities.

Nutritional Value.

Improvements sought for existing crops went beyond their growing characteristics and their resistance to natural enemies. Research was also directed at modifying the protein structure of the crops by altering the amino acid chain in the plant's DNA. Among the expected benefits of this research is the enhancement of the usable protein in staple crops like corn and wheat. In their 1983 article in Science, Barton and Brill outlined the possibilities.

The seeds of legumes and cereal grains provide humans directly with approximately 70 per cent of their dietary protein requirement. Throughout seed development, storage proteins are synthesized and accumulated within the seed, apparently to provide a source of amino acid reserves during early seed germination.

High levels of such protein in seeds provides an enriched amino acid source for both human and animal consumption. However, various deficiencies of seeds in certain essential amino acids do not allow either cereal grains or legumes to provide a balanced diet without supplementation of the limiting amino acids from other sources. Thus, one approach towards the problem of malnutrition is the improvement of nature's "nutritionally deficient seeds" by encoding them with useful storage proteins. Molecular genetics is viewed as a tool for creating perfect or nearly perfect food sources–foods that carry a complete nutritional balance of amino acids.

For lesser developed countries that must import food products to round out the nutritional complement of indigenous crops, or for nations where there has been a scarcity of certain grains, this could be a great asset. But would this technology empower the lesser developed countries? Or would it create a new form of domination by the industrialized nations–a kind of genetic colonialism?

Herbicide Resistance.

Nature did not endow most plants with a resistance to synthetic herbicides. Recently, with the extravagant use of herbicides in agriculture and in war as defoliants, plants have been discovered with natural resistance to herbicides. In 1982, herbicides represented 47 per cent of the $3.7 billion a year U.S. pesticide market, eclipsing insecticides and fungicides. Biotechnology offered the herbicide industry a way to expand markets and at the same time create a growing environment that simulated a weed-free green house.

By engineering crops with herbicide-resistant traits, the herbicide could be used at any stage in the growing cycle. This so-called improvement upon nature does not take into consideration the effects of the increased use of the herbicide. Guido Ruivenkamp, a Dutch researcher at the University of Amsterdam, wrote that "big industry may see the early opportunity of developing pesticide-resistant plants rather than undertaking the longer term effort of developing pest-resistant plants."

Plants that Grow Faster.

A report issued by the National Academy of Sciences titled Biotechnology: An Industry Comes of Age describes an enzyme with the acronym Rubisco that is an important catalyst in photosynthesis. If scientists learn how to move the genes for Rubisco it may be possible to amplify the genes' function and accelerate plant growth. An obvious advantage to the farmer would be the addition of another growth cycle, making more efficient use of agricultural land.

MODIFICATION OF ANIMALS

Transgenics is the term used to describe the application of rDNA techniques to genetically modify animal species. In some instances, the current research strategy represents an extension of traditional goals to create animals that meet certain requirements of the consumer (as interpreted by agribusiness) or of the food processing industry. For example, there is an interest in creating livestock with leaner meat, and in producing breastier fowl. Since these animals are used exclusively in food production they are not suited to survive anywhere beyond the feedlot. As such they are unlikely to pose environmental problems from a release. The genetic redesign of animal species has other applications which might affect the environment should the animals be released. There are research programmes to use transgenic animals as functioning, living bioreactors. For example, an animal may be genetically engineered to secrete a human protein in its milk or blood stream. In 1987 a transgenic mouse was produced that secreted a clot-dissolving drug in its milk.

Mice are produced twice their normal size by the endowment of a human growth gene. In yet another case, the gene that codes for the AIDS virus was implanted in the fertilized egg of a mouse and brought to term. The AIDS gene is found (albeit not necessarily expressed) in every cell of the mouse.

Release of such a mouse into the environment could represent a new route for the spread of the disease. It is for this reason that NIH took special precautions in handling these experimental mice by keeping them in high-containment facilities. There has also been experimentation with transgenic fish to determine if they can be genetically designed to survive in colder water or if a larger species can be introduced to repopulate a river. The lingering question is: what impact would such releases have on the aquatic ecology?

MINING OF THE EARTH

Oil Recovery. Microorganisms can be used to enhance recovery methods for removing crude oil from wells. Initial research suggested that certain organisms can reduce oil viscosity, improve rock porosity and permeability, and decrease water-oil interfacial tension. The technique of Microbiologically Enhanced Oil Recovery (MEOR) was first developed in the 1940s. Through genetic engineering, there was a new impetus to improve MEOR by enhancing the traits of organisms like *Clostridium acetobatylicum,* which has been used in the extraction of oil.

The microorganism *Xanthomonas campestris,* which produces xanthum gum, is used for oil recovery by increasing the viscosity of water that is pumped into wells to displace the oil. Another MEOR approach that claims success in recovering oil from marginal wells creates a mixture of microbes and molasses which is injected into oil wells. The gas produced by the microbial growth increases the pressure, forcing out the oil.

Microbial Mining

Microorganisms have also been useful in recovering certain metals from ore-bearing minerals. Copper and uranium are mined and commercially purified by a process that includes leaching by bacteria such as *Thiobacillus ferroxidans.* The leaching process may be done in situ or in a controlled environment. Efforts are underway to improve the leaching properties of the organisms through recombinant DNA techniques.

The authors offer three examples:

- Bacterial leaching operations that generate large quantities of sulfuric acid could acidify water supplies;

- *Thiobacilli* and related species may acquire the traits to infect humans;
- Metals concentrated by bacteria from dilute mine waters can accumulate in the food chain.

BIOSENSORS

Microorganisms may also be used as markers in the environment. The concerns about releasing novel organisms have produced an interest in the methods of monitoring the dispersal of the agents. Monsanto Company has developed a genetically engineered strain of *Pseudomonas* carrying natural resistance to antibiotics and genes from *E. coli* that code for proteins that enable the microbe to metabolize lactose. These traits make the organism a possible marker for field tests. The first approval by the EPA of a field test of a genetically engineered organism under the Toxic Substances Control Act (TSCA) was for Monsanto's genetically marked microbe. The test was performed in Blacksville, South Carolina.

The environmental uses of applied genetics constitute an important part of the industrial agenda. But the prospects of releasing genetically novel plants, microbes, and animals have activated a lively public policy debate.

The nature and origins of these concerns and shows the regulatory transition that took place from laboratory experiments to field experiments. The next generation of environmental experiments and applications of biotechnics brought a new group of scientist-stakeholders into the discourse of risks and benefits of new technologies.

3

The Structure of Genomes

The most significant breakthroughs in agricultural biotechnology are coming from research into the structure of genomes and the genetic mechanisms behind economically important traits.

The rapidly progressing discipline of genomics is providing information on the identity, location, impact and function of genes affecting such traits - knowledge that will increasingly drive the application of biotechnology in all agricultural sectors.

Genomics sets the foundation for post-genomics activities, including new disciplines such as proteomics and metabolomics to generate knowledge on gene and protein structure, as well as their functions and interactions. These disciplines seek to understand systematically the molecular biology of organisms for their practical use. A vast range of new and rapidly advancing technologies and equipment has also been developed to generate and process information about the structure and function of biological systems.

The use and organization of this information is called bioinformatics. Advances in bioinformatics may allow the prediction of gene function from gene sequence data: from a listing of an organism's genes, it will become possible to build a theoretical framework of its biology. The comparison across organisms of physical and genetic maps and DNA sequences will significantly reduce the time needed to identify and select potentially useful genes. Through the production of genetic maps that provide the precise location and sequences of genes, it is apparent that even distantly related genomes share common features. Comparative genomics assists in the understanding of many genomes based on the intensive study of just a few. For instance, the rice genome sequence is useful for studying the genomes of other cereals with which it shares features according to its degree of relatedness, and the mouse and malaria genomes provide models for livestock and some of the diseases that affect them. There are now model species for most types of crops, livestock and diseases and knowledge of their genomes is accumulating rapidly.

STRUCTURE OF DNA IN THE GENETIC CODE

The genetic code consists of 64 triplets of nucleotides. These triplets are called codons.With three exceptions, each codon encodes for one of the 20 amino acids used in the synthesis of proteins. That produces some redundancy in the code: most of the amino acids being encoded by more than one codon. The structure of DNA encodes all the information every cell needs to function and thrive.

In addition, DNA carries hereditary information in a form that can be copied and passed intact from generation to generation. A gene is a segment of DNA.

The biochemical instructions found within most genes, known as the genetic code, specify the chemical structure of a particular protein. Proteins are composed of long chains of amino acids, and the specific sequence of these amino acids dictates the function of each protein. The DNA structure of a gene determines the arrangement of amino acids in a protein, ultimately determining the type and function of the protein manufactured.

DNA STRUCTURE

DNA molecules form from chains of building blocks called nucleotides. Each nucleotide consists of a sugar molecule called deoxyribose that bonds to a phosphate molecule and to a nitrogen-containing compound, known as a base. DNA uses four bases in its structure: adenine (A), cytosine (C), guanine (G), and thymine (T).

The order of the bases in a DNA molecule—the genetic code—determines the amino acid sequence of a protein. In the cells of most organisms, two long strands of DNA join in a single molecule that resembles a spiraling ladder, commonly called a double helix.

Alternating phosphate and sugar molecules form each side of this ladder. Bases from one DNA strand join with bases from another strand to form the rungs of the ladder, holding the double helix together. The pairing of bases in the DNA double helix is highly specific—adenine always joins with thymine, and guanine always links to cytosine. These base combinations, known as complementary base pairing, play a fundamental role in DNA's function by aiding in the replication and storage of genetic information.

Complementary base pairing also enables scientists to predict the sequence of bases on one strand of a DNA molecule if they know the order on the corresponding, or complementary, DNA strand. Scientists use complementary base pairing to help identify the genes on a particular chromosome and to develop methods used in genetic engineering. Genes line up in a row along the length of a DNA molecule.

In humans a single gene can vary in length from 100 to over 1,000,000 bases. Genes make up less than 2 percent of the length of a DNA molecule. The rest of the DNA molecule is made up of long, highly repetitive nucleotide sequences. Once dismissed as "junk" DNA, scientists now believe these nucleotide sequences may play a role in the survival of cells. Identifying the function of these sequences is a thriving field of genetics research.

REPLICATION OF DNA

DNA replication begins with the "unzipping" of the parent molecule as the hydrogen bonds between the base pairs are broken. In order for inherited traits to be transmitted from parent to child, the genetic information encoded in DNA must be copied with great precision during cell division. The accuracy of DNA replication depends upon the complementary pairing of bases.

During replication, the DNA double helix unwinds and bonds joining the base pairs break, separating the DNA molecule into two separate strands. Each strand of DNA directs the synthesis of another complementary strand. The unpaired bases of each DNA strand attach to bases floating within the cell. But the DNA strand's unpaired bases bond only with specific, complementary bases—for example, an adenine base will bond only with a thymine base and a cytosine bases will pair only with a guanine base. Once all of the bases of a DNA strand bond to complementary bases, the complementary bases then link to each other, forming a new DNA double-helix molecule. Thus the original DNA molecule replicates into two DNA molecules that are exact duplicates.

PROTEIN SYNTHESIS IN DNA REPLICATION

All organisms must duplicate their DNA with extraordinary accuracy before each cell division. In this section, we explore how an elaborate "replication machine" achieves this accuracy, while duplicating DNA at rates as high as 1000 nucleotides per second. DNA replication ensures that the genetic instructions encoded in DNA can be used continuously through generations to produce the proteins that build and operate the cells of an organism. The process of tapping the genetic code to create proteins, known as protein synthesis, has two crucial steps: transcription and translation.

Transcription transfers of the genetic code: Transcription transfers the genetic code from a molecule of DNA to an intermediary molecule called ribonucleic acid (RNA). The basic nucleotide structure of RNA resembles that of DNA, but the two compounds have three critical differences.

First, the structure of RNA incorporates the sugar ribose rather than deoxyribose, the sugar in DNA.

Second, RNA uses the base uracil (U) instead of thymine (T). In RNA uracil binds with adenine just as thymine does in DNA.

Third, RNA usually exists as a single strand, unlike the double-helix structure that normally characterizes DNA.

Transcription involves the production of a special kind of RNA known as messenger RNA (mRNA). The process begins when the two strands of a DNA molecule separate, a task directed by the enzyme RNA polymerase. After the double helix splits apart, one of the strands serves as a template, or pattern, for the formation of a complementary mRNA molecule.

Free-floating individual bases within the cell bind to the bases on the DNA template using complementary base pairing. The individual bases then link together to form a strand of mRNA. In eukaryotes (organisms whose cells have a nucleus), the mRNA strand undergoes an additional step before the next stage of protein synthesis can occur. The mRNA strand consists of coding regions called exons separated by regions called introns.

The introns do not contribute to protein synthesis. Special enzymes in the nucleus remove the introns from the mRNA strand. The remaining exons then link together to form an mRNA strand that contains the entire code for making a protein.

Translation of the genetic code: Once transcription is complete and the genetic code has been copied onto mRNA, the genetic code must be converted into the language of proteins. That is, the information coded in the four bases found in mRNA must be translated into the instructions encoded by the 20 amino acids used in the formation of proteins. This process, called translation, takes place in cellular organelles called ribosomes.

In eukaryotes, mRNA travels out of the nucleus into the cell body to attach to a ribosome. In prokaryotes (organisms without a nucleus), the ribosome clasps mRNA and starts translation before these strands have finished transcription and separated from the DNA. In both eukaryotes and prokaryotes, the ribosome acts like a workbench and clamp that holds the mRNA strand and coordinates the activity of enzymes and other molecules essential to translation. Another form of RNA called transfer RNA (tRNA) is found in the cytoplasm of the cell.

There are many different types of tRNA, and each type binds with one of the 20 amino acids used in protein formation. One end of a tRNA binds with a specific amino acid. The other end carries three bases, known as an anticodon. The tRNA with an amino acid attached travels to the ribosome where the mRNA is stationed. The anticodon of the tRNA undergoes complementary base pairing with a series of three bases on the mRNA, known as the codon. The mRNA codon codes for the type of amino acid carried by the tRNA. A second tRNA bonds with the next codon on the mRNA.

The resident tRNA transfers its amino acid to the amino acid of the incoming tRNA and then leaves the ribosome. This process continues repeatedly, with new tRNA receiving the growing chain of amino acids, known as a polypeptide chain, from a resident tRNA. The ribosome moves the mRNA strand one codon at a time, making new codons available to bind with tRNAs. The process ends when the entire sequence of mRNA has been translated. The polypeptide chain falls away from the ribosome as a newly formed protein, ready to go to work in the cell.

MUTATIONS OF DNA REPLICATION AND PROTEIN SYNTHESIS

Occasionally mistakes occur during DNA replication and protein synthesis. Any alteration in the structure of a gene results in a mutation. Mutations occur during DNA replication when the chemical structure of genes undergoes random modifications.

Once a change has occurred, the altered genes continue to replicate in their changed form unless another mutation occurs. Sometimes mutations occur during transcription or translation, causing protein synthesis to go awry. Although mutations may occur in any living cell, they are most important when they occur in gametes because then the change affects the traits of following generations. Most mutations harm an organism.

If a mutation occurs in a gene sequence that codes for a particular protein, the mutation may result in a change in the amino acid sequence directed by the gene. This change, in turn, may affect the function of the protein. The implications can be significant: The amino acid sequence distinguishing normal hemoglobin from the altered form of hemoglobin responsible for sickle-cell anemia differs by only a single amino acid. Some mutations may be neutral or silent and do not affect the function of a protein.

Occasionally a mutation benefits an organism. Over the course of evolutionary time, however, mutations serve the crucial role of providing organisms with previously non-existent proteins. In this way, mutations are a driving force behind genetic diversity and the rise of new or more competitive species better able to adapt to changes, such as climate variations, depletion of food sources, or the emergence of new types of disease. Mutations can produce a change in any region of a DNA molecule. In a *point mutation*, for example, a single nucleotide replaces another nucleotide.

Although a point mutation produces a small change to the DNA sequence, it may cause a change in the amino acid sequence, and thus the function, of a protein. Far more serious are mutations that involve the addition or deletion of one or more bases from a DNA molecule. Adding or subtracting even a single base from a normal sequence during transcription can disrupt translation by shifting the "reading frame" of every subsequent codon.

For example, an mRNA strand may include two codons in the following sequence: AUG UGA. The addition of a cytosine base at the beginning of this sequence shifts the "spelling" of these codons so that they read: CAU GUG. This may result in an incorrect amino acid sequence during translation, or the protein may be truncated. Known as *frameshift mutation*, this type of alteration could result in the production of a protein with no real function or one with a harmful effect. Sometimes mutations are caused by *transposition*, in which long stretches of DNA (containing one or more genes) move from one chromosome to another.

These jumping genes, called transposons, can disrupt transcription and change the type of amino acids inserted into a protein. Transposons rearrange and interrupt genes in a way that generally improves the genetic variation of a species. While mutations can occur spontaneously, some can be caused by exposure to physical or chemical agents in the environment called mutagens. Common environmental mutagens include ultraviolet rays from the sun and various chemicals, such as asbestos, cigarette smoke, and nitrous acid. High-energy radiation, such as medical X rays, can cause DNA strands to break, leading to the deletion of potentially important genetic information. Radiation damage can also affect an entire chromosome, disrupting the function of many genes.

In chromosomal translocation, a piece of one chromosome breaks off and merges with another chromosome. In some cases, large sections of chromosomes may break off and be lost. The cell has highly effective self-repair mechanisms that can correct the harmful changes made by mutations and prevent some mutations from being passed on. Some 50 specialized enzymes locate different types of faulty sequences in the DNA and clip out those flaws. Another repair mechanism scans DNA after replication and marks mismatched base pairs for repair.

SMALL FRACTION OF THE DNA

All living things are made up of cells that are programmed by genetic material called deoxyribonucleic acid (DNA). Only a small fraction of the DNA chain actually makes up genes, which in turn code for proteins, and the remaining share of the DNA represents non-coding sequences whose role is not yet clearly understood.

The genetic material is organized into pairs of chromosomes. For example, there are five chromosome pairs in the much-studied mustard species Arabidopsis thaliana. An organism's entire set of chromosomes is called the genome. The Human Genome Sequencing Project has provided the agricultural research community not only with many spin-off technologies that can be applied across the board for all living organisms but also with a model for international collaboration in tackling large genome-sequencing projects for model plants such as Arabidopsis and rice. Synteny describes the conservation or consistency of gene content and gene order along the chromosomes of different plant genomes.

Until well into the 1980s we imagined that each crop plant had its own genetic map. Only when we were able to make the first molecular maps, using a technique called "restriction fragment length polymorphism" (RFLP), did it begin to dawn on us that related species had remarkably similar gene maps. The early experiments demonstrated conservation over a few million years of evolution in syntenous relationships between potato and tomato in the broad-leafed plants and between the three genomes of bread wheat in the grasses.

Later we were able to show that the same similarities held over the rice, wheat and maize genomes, which were separated by some 60 million years of evolution. The diagram summarizes this research and shows 70 per cent of the world's food linked in a single map. The 12 chromosomes of rice can be aligned with the ten chromosomes of maize and the basic seven chromosomes of wheat and barley in such a way that any radius drawn around the circles will pass through different versions, known as alleles, of the same genes. The discovery of synteny has had an enormous impact on the way we think about plant genetics.

There are obvious applications for evolutionary studies; for example, the white arrows on the wheat and maize circles describe evolutionary chromosomal translocations that describe *Pooideae* and *Panicoideae* groups of grasses. There are great opportunities to predict the presence and location of a gene in one species from what we know from another. Now that we have the complete DNA sequence of rice we are able to identify and isolate key genes from large genome intractable species such as wheat and barley by predicting that the same genes will be present in the same order as in rice. Key genes for disease resistance and tolerance to acid soils have recently been isolated from barley and rye in this way. For practical plant breeding, knowledge of synteny allows breeders access to all alleles in, for example, all cereals rather than just the species on which they are working.

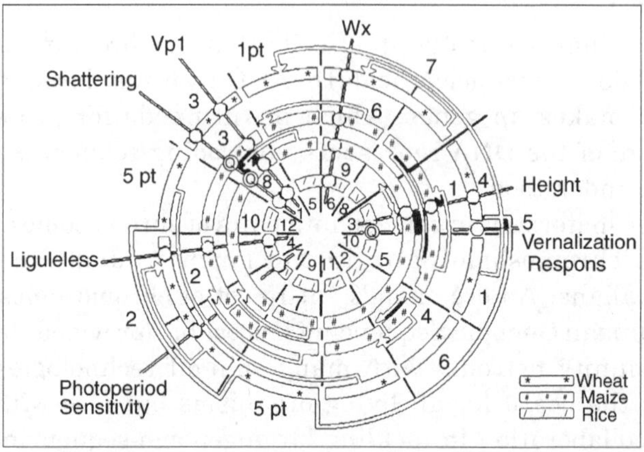

A key first example of this is the transfer to rice of the wheat dwarfing genes that made the Green Revolution possible. In these experiments the gene was located in rice by synteny and then isolated and engineered with the alteration in DNA sequence that characterized the wheat genes before replacing the engineered gene in rice. This approach can be applied to any gene in any cereal, including the so-called "orphan crops" that have not attracted the research dollars that the big three - wheat, rice and maize - have over the past century. The main significance is, however, that we can now pool our knowledge of biochemistry, physiology and genetics and transfer it between crops via synteny.

STRUCTURE OF DNA

DNA stands for deoxyribonucleic acid. DNA is pretty unusual in that it is about the only common molecule capable of directing its own synthesis. The processes of mitosis and meiosis were discovered in the 1870s and 1890s. It was observed that, as cells divided, chromosomes moved around in a cell, and people began to wonder what their function was. It was determined that chromosomes were made of protein and DNA, about which people knew almost nothing. People began to suspect that chromosomes had something to do with genetics, but couldn't explain what/how. When enough evidence was accumulated to confirm that chromosomes did, indeed, have something to do with genetics, most people thought that in some way the protein in the chromosomes served as the genetic material. People knew that DNA was also in the chromosomes, but because its structure was unknown and people didn't know much about it, few people thought it was the genetic material.

The names of James Watson and Francis Crick are so closely linked with DNA that it is easy to forget that, when they began their collaboration in Cambridge, England in October 1951, the detailed structure of the DNA polymer was already known. Their contribution was not to determine the structure of DNA *per se*, but to show that in living cells two DNA chains are intertwined to form the double helix. We will consider the two facets of DNA structure separately.

NUCLEOTIDES AND POLYNUCLEOTIDES

DNA is a linear, unbranched polymer in which the monomeric subunits are four chemically distinct nucleotides that can be linked together in any order in chains hundreds, thousands or even millions of units in length. Each nucleotide in a DNA polymer is made up of three components:

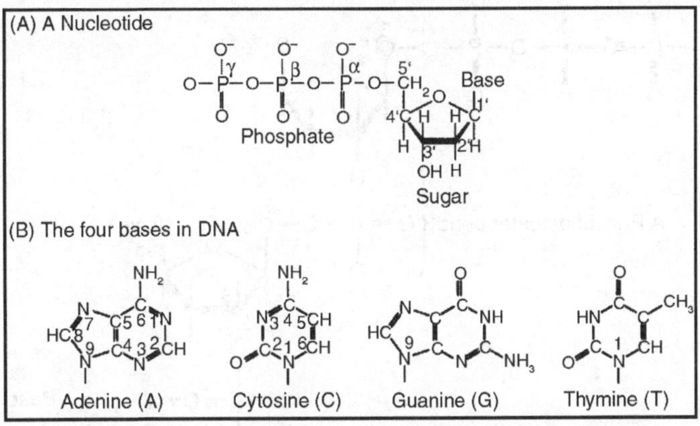

Fig. The Structure of a Nucleotide. (A) The General Structure of a Deoxyribonucleotide, the type of Nucleotide found in DNA. (B) The four Bases that Occur in Deoxyribonucleotides.

1. 2-deoxyribose, which is a pentose, a type of sugar composed of five carbon atoms. These five carbons are numbered 1' (spoken as 'one-prime'), 2, etc. The name '2-deoxyribose' indicates that this particular sugar is a derivative of ribose, one in which the hydroxyl (-OH) group attached to the 2-carbon of ribose has been replaced by a hydrogen (-H) group.
2. A nitrogenous base, one of cytosine, thymine (single-ring pyrimidines), adenine or guanine (double-ring purines). The base is attached to the 1-carbon of the sugar by a b-*N*-glycosidic bond attached to nitrogen number 1 of the pyrimidine or number 9 of the purine.

3. A phosphate group, comprising one, two or three linked phosphate units attached to the 5-carbon of the sugar. The phosphates are designated a, b and g, with the a-phosphate being the one directly attached to the sugar.

A molecule made up of just the sugar and base is called a nucleoside; addition of the phosphates converts this to a nucleotide. Although cells contain nucleotides with one, two or three phosphate groups, only the nucleoside triphosphates act as substrates for DNA synthesis.

The full chemical names of the four nucleotides that polymerize to make DNA are:

1. 2'-deoxyadenosine 5-triphosphate
2. 2'-deoxycytidine 5-triphosphate
3. 2'-deoxyguanosine 5-triphosphate
4. 2'-deoxythymidine 5-triphosphate

The abbreviations of these four nucleotides are dATP, dCTP, dGTP and dTTP, respectively, or, when referring to a DNA sequence, A, C, G and T, respectively.

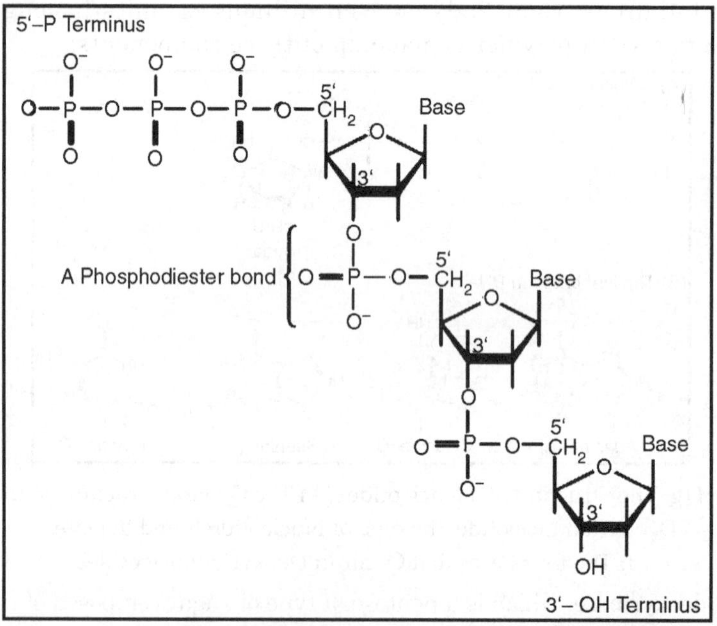

Fig. A Short DNA Polynucleotide Showing the Structure of the Phosphodiester bond.

In a polynucleotide, individual nucleotides are linked together by phosphodiester bonds between their 5- and 3-carbons. From the structure of this linkage we can see that the polymerization reaction involves removal of the two outer phosphates (the β- and γ-phosphates) from one nucleotide and replacement of the hydroxyl group attached to the 3-carbon of the second nucleotide. Note that the two ends of the polynucleotide are chemically distinct, one having an unreacted triphosphate group attached to the 5-carbon (the 5 or 5-P terminus) and the other having an unreacted hydroxyl attached to the 3-carbon (the 3 or 3-OH terminus). An important consequence of the polarity of the phosphodiester bond is that the chemical reaction needed to extend a DNA polymer in the 5→3 direction is different to that needed to make a 3→5 extension. All natural DNA polymerase enzymes are only able to carry out 5→3 synthesis, which adds significant complications to the process by which double-stranded DNA is replicated. The same limitation applies to RNA polymerases, the enzymes which make RNA copies of DNA molecules.

RNA

Although our attention is firmly on DNA, the structure of RNA is so similar to that of DNA that it makes sense to introduce it here. RNA is also a polynucleotide but with two differences compared with DNA. First, the sugar in an RNA nucleotide is ribose and, second, RNA contains uracil instead of thymine.

The four nucleotide substrates for synthesis of RNA are therefore:

1. Adenosine 5-triphosphate
2. Cytidine 5-triphosphate
3. Guanosine 5-triphosphate
4. Uridine 5-triphosphate

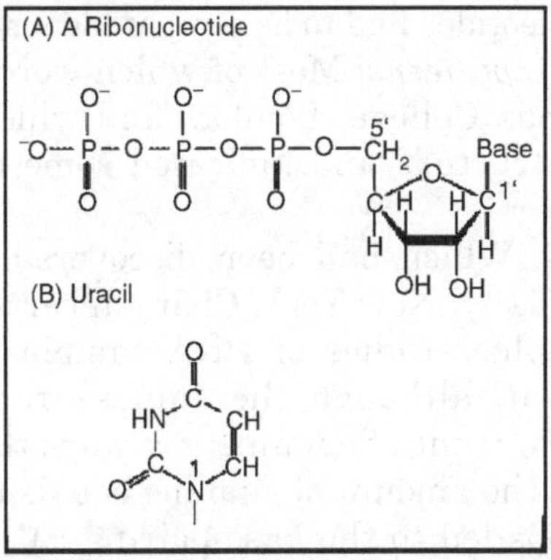

Fig. The Chemical Differences Between DNA and RNA.

(A) RNA Contains Ribonucleotides, in which the Sugar is Ribose Rather than 2'-deoxyribose. The difference is that a Hydroxyl group Rather than Hydrogen atom is Attached to the 2-carbon. (B) RNA Contains the Pyrimidine called Uracil Instead of Thymine which are Abbreviated to ATP, CTP, GTP and UTP, or A, C, G and U, Respectively. As with DNA, RNA polynucleotides contain 3_5 phosphodiester bonds, but these phosphodiester bonds are less stable than those in a DNA polynucleotide because of the indirect effect of the hydroxyl group at the 2-position of the sugar. This may be one reason why the biological functions of RNA do not require the polynucleotide to be more than a few thousand nucleotides in length, at most. There are no RNA counterparts of the million-unit sized DNA molecules found in human chromosomes.

THE DOUBLE HELIX

In the years before 1950, various lines of evidence had shown that cellular DNA molecules are comprised of two or more polynucleotides assembled together in some way. The possibility that unraveling the nature of this assembly might provide insights into how genes work prompted Watson and Crick, among others, to try to solve the structure. According to Watson in his book *The Double Helix*, their work was a desperate race against the famous American biochemist, Linus Pauling, who initially proposed an incorrect triple helix model, giving Watson and Crick the time they needed to complete the double helix structure. It is now difficult to separate fact from fiction, especially regarding the part played by Rosalind Franklin, whose X-ray diffraction studies provided the bulk of the experimental data in support of the double helix and who was herself very close to solving the structure.

The Evidence that Led to the Double Helix

Watson and Crick used four types of information to deduce the double helix structure:

- *Biophysical data of various kinds*: The water content of DNA fibres was particularly important because it enabled the density of the DNA in a fibre to be estimated. The number of strands in the helix and the spacing between the nucleotides had to be compatible with the fibre density.
- *X-ray diffraction patterns*: Most of which were produced by Rosalind Franklin of Kings College, London, and which revealed the helical nature of the structure and indicated some of the key dimensions within the helix.
- *The base ratios*: Which had been discovered by Erwin Chargaff of Columbia University, New York. Chargaff carried out a lengthy series of chromatographic studies of DNA samples from various sources and showed that, although the values are different in different organisms, the amount of adenine is always the same as the amount of thymine, and the amount of guanine equals the amount of cytosine. These base ratios led to the base-pairing rules, which were the key to the discovery of the double helix structure.

DNA was Extracted from Various Organisms and Treated with Acid to Hydrolyze the Phosphodiester Bonds and Release the Individual Nucleotides. Each Nucleotide was then Quantified by Chromatography. The data show some of the Actual results Obtained by Chargaff. These Indicate that, within Experimental Error, the Amount of Adenine is the same as that of Thymine, and the Amount of Guanine is the same as that of Cytosine.

- *Model building*: Which was the only major technique that Watson and Crick made use of themselves. Scale models of possible DNA structures enabled the relative positioning of the various atoms to be checked, to ensure that pairs of groups that formed bonds were not too far apart, and that other groups were not so close together as to interfere with one another.

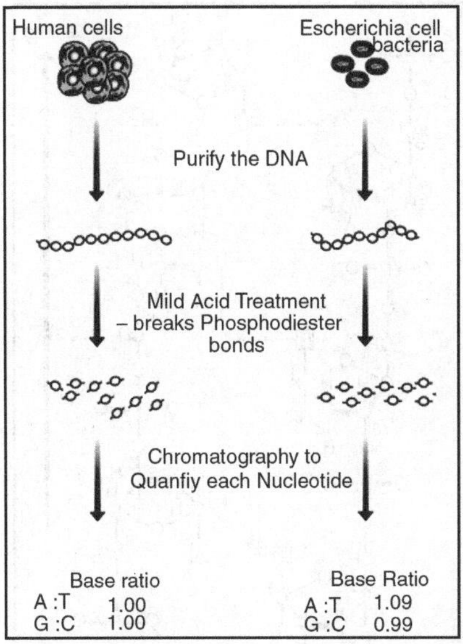

Fig. The Base Ratio Experiments Performed by Chargaff.

The Key Features of the Double Helix

The double helix is right-handed, which means that if it were a spiral staircase and you were climbing upwards then the rail on the outside of the staircase would be on your right-hand side. The two strands run in opposite directions. The helix is stabilized by two types of chemical interaction:

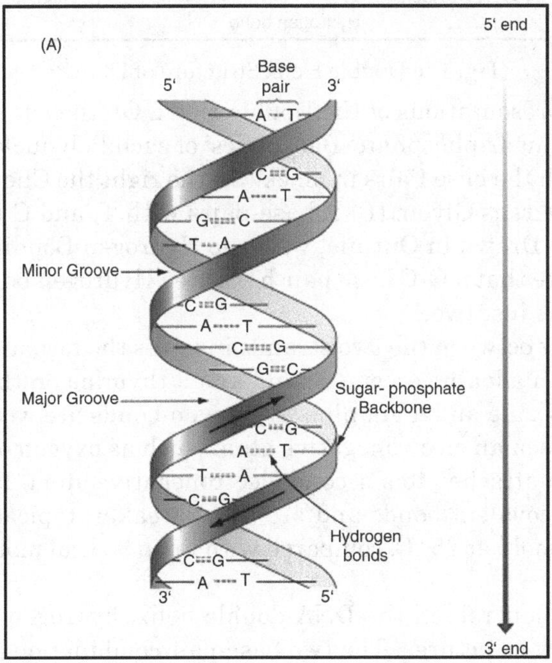

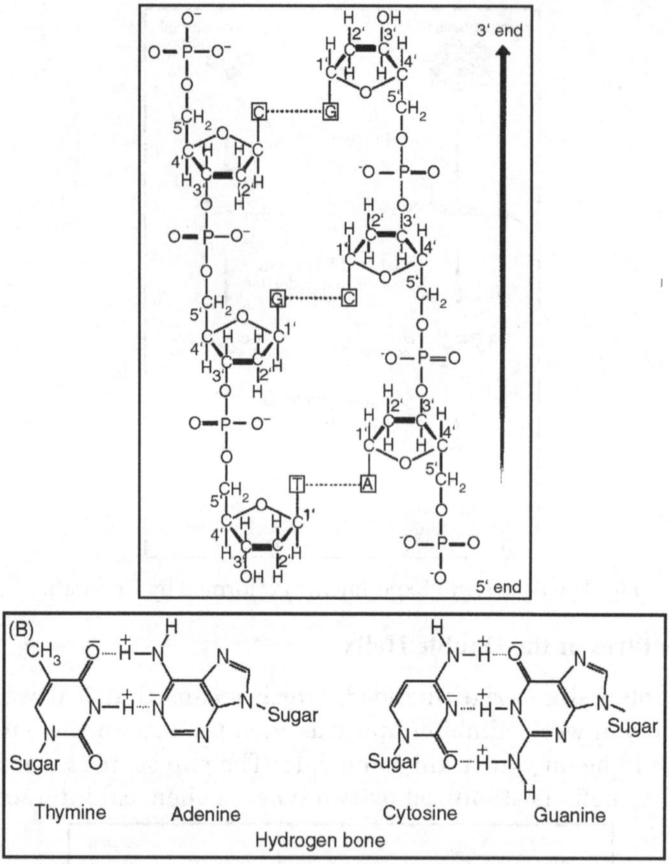

Fig. The Double Helix Structure of DNA.

(A) Two Representations of the Double Helix. On the left the Structure is shown with the Sugar-phosphate 'Backbones' of each Polynucleotide drawn as a red Ribbon with the base Pairs in black. On the right the Chemical Structure for three base Pairs is Given. (B) A base-pairs with T, and G base-pairs with C. The bases are Drawn in Outline, with the Hydrogen Bonding Indicated by Dotted lines. Note that a G-C base pair has three Hydrogen bonds whereas an A-T base pair has just two.

Base-pairing between the two strands involves the formation of hydrogen bonds between an adenine on one strand and a thymine on the other strand, or between a cytosine and a guanine. Hydrogen bonds are weak electrostatic attractions between an electronegative atom (such as oxygen or nitrogen) and a hydrogen atom attached to a second electronegative atom. Hydrogen bonds are longer than covalent bonds and are much weaker, typical bond energies being 1_10 kcal mol^{-1} at 25 °C, compared with up to 90 kcal mol^{-1} for a covalent bond.

As well as their role in the DNA double helix, hydrogen bonds stabilize protein secondary structures. The two base-pair combinations- A base-paired with T, and G base-paired with C - explain the base ratios discovered by Chargaff. These are the only pairs that are permissible, partly because of the geometries of the nucleotide bases and the relative positions of the groups that are able to participate in hydrogen bonds, and partly because the pair must be between a purine and a pyrimidine; a purine-purine pair would be too big to fit within the helix, and a pyrimidine-pyrimidine pair would be too small.

- Base-stacking, sometimes called p-p interactions, involves hydrophobic interactions between adjacent base pairs and adds stability to the double helix once the strands have been brought together by base-pairing. These hydrophobic interactions arise because the hydrogen-bonded structure of water forces hydrophobic groups into the internal parts of a molecule.

Both base-pairing and base-stacking are important in holding the two polynucleotides together, but base-pairing has added significance because of its biological implications. The limitation that A can only base-pair with T, and G can only base-pair with C, means that DNA replication can result in perfect copies of a parent molecule through the simple expedient of using the sequences of the pre-existing strands to dictate the sequences of the new strands.

This is template-dependent DNA synthesis and it is the system used by all cellular DNA polymerases. Its counterpart, template-dependent RNA synthesis, is used by RNA polymerases to make RNA copies of genes, these copies preserving the biological information contained in the sequence of the genomic DNA molecule.

The only difference between DNA and RNA syntheses is that when RNA is made, the adenines in the DNA template do not specify thymines in the RNA copy. This is because RNA does not contain thymine; instead z pairs with uracil in DNA-RNA hybrids and in double-stranded RNA structures.

The Double Helix has Structural Flexibility

The double helix described by Watson and Crick, and shown in *Figure*, is called the B-form of DNA. Its characteristic features lie in its dimensions: a helical diametre of 2.37 nm, a rise of 0.34 nm per base pair, and a pitch (*i.e.*, distance taken up by a complete turn of the helix) of 3.4 nm, this corresponding to ten base pairs per turn.

The DNA in living cells is thought to be predominantly in this B-form, but it is now clear that genomic DNA molecules are not entirely uniform in structure. This is mainly because each nucleotide in the helix has the flexibility to take up slightly different molecular shapes. To adopt these different conformations, the relative positions of the atoms in the nucleotide must change slightly. There are a number of possibilities but the most important conformational changes involve rotation around the β-*N*-glycosidic bond, changing the orientation of the base relative to the sugar, and rotation around the bond between the 3′- and 4′-carbons.

Both rotations have a significant effect on the double helix: changing the base orientation influences the relative positioning of the two polynucleotides, and rotation around the 3_4 bond affects the conformation of the sugar-phosphate backbone. Rotations within individual nucleotides therefore lead to major changes in the overall structure of the helix. It has been recognized since the 1950s that changes in the dimensions of the double helix occur when fibres containing DNA molecules are exposed to different relative humidities. For example, the modified version of the double helix called the A-form has a diametre of 2.55 nm, a rise of 0.29 nm per base pair and a pitch of 3.2 nm, corresponding to 11 base pairs per turn. Other variations include B-, C-, C-, C-, D-, E- and T-DNAs. All these are right-handed helices like the B-form.

Table. Features of Different Conformations of the DNA Double Helix

Feature	B-DNA	Conformation A-DNA	Z-DNA
Type of helix	Right-handed	Right-handed	Left-handed
Helical diameter (nm)	2.37	2.55	1.84
Rise per base pair (nm)	0.34	0.29	0.37
Distance per complete turn (pitch) (nm)	3.4	3.2	4.5
Number of base pairs per complete turn	10	11	12
Topology of major groove	Wide, deep	Narrow, deep	Flat
Topology of minor groove	Narrow, shallow	Broad, shallow	Narrow, deep

A more drastic reorganization is also possible, leading to the left-handed Z-DNA, a slimmer version of the double helix with a diametre of only 1.84 nm. The bare dimensions of the various forms of the double helix do not reveal what are probably the most significant differences between them. These relate not to diametre and pitch, but the extent to which the internal regions of the helix are accessible from the surface of the structure.

As shown in and. DNA does not have an entirely smooth surface; instead, two grooves spiral along the length of the helix. One of these grooves is relatively wide and deep and is called the major groove; the other is narrow and less deep and is called the minor groove. A-DNA also has two grooves but with this conformation the major groove is even deeper, and the minor groove shallower and broader. Z-DNA is different again, with one groove virtually non-existent but the other very narrow and deep.

In each form of DNA part of the internal surface of at least one of the grooves is formed by chemical groups attached to the nucleotide bases. we will see that expression of the biological information contained within a genome is mediated by DNA-binding proteins which attach to the double helix and regulate the activity of the genes contained within it. To carry out their function, each DNA-binding protein must attach at a specific position, near to the gene whose activity it must influence. This can be achieved, with a greater or lesser degree of ambiguity, by the protein reaching down into a groove, within which the DNA sequence can be 'read' without the helix being opened up by breaking the base pairs.

THE DISTRIBUTION OF GENETIC VARIATION

Reliable information on the distribution of genetic variation is a prerequisite for sound selection, breeding and conservation programmes. Genetic variation of a species or population can be assessed in the field or by studying molecular and other markers in the laboratory.

A combination of the two approaches is required for reliable results. Molecular markers are identifiable DNA sequences, found at specific locations of the genome and associated with the inheritance of a trait or linked gene.

Molecular markers can be used for:

- Marker-assisted breeding,
- Understanding and conserving genetic resources
- Genotype verification. These activities are critical for the genetic improvement of crops, forest trees, livestock and fish.

GENES AFFECTING TRAITS OF ECONOMIC IMPORTANCE IN PLANTS OR ANIMALS

Genetic linkage maps can be used to locate and select for genes affecting traits of economic importance in plants or animals. The potential benefits of marker-assisted selection (MAS) are greatest for traits that are controlled by many genes, such as fruit yield, wood quality, disease resistance, milk and meat production, or body fat, and that are difficult, time-consuming or expensive to measure.

Markers can also be used to increase the speed or efficiency of introducing new genes from one population to another, for example when wishing to introduce genes from wild relatives into modern plant varieties. When the desired trait is found within the same species, it may be transferred with traditional breeding methods, with molecular markers being used to track the desired gene. Pearl millet is a cereal grown for foodgrain and straw in the hottest, driest areas of Africa and Asia where rainfed and dryland agriculture are practised.

It is similar to maize in its breeding behaviour. Traditional farmers' varieties are open-pollinated and out-breeding and thus continuously changing. Genetically uniform hybrid varieties have been developed that offer higher yield potential but are more vulnerable to a plant disease called downy mildew. In India, pearl millet is grown on about 9 million ha and more than 70 per cent of this is sown to such hybrid cultivars. Since pearl millet hybrids first reached farmers' fields in India in the late 1960s, every variety that has become popular with farmers has ultimately succumbed to a downy mildew epidemic.

Unfortunately, by the time the poorer farmers in a given region decide to adopt a particular variety, its days are usually numbered. The International

Crops Research Institute for the Semi-Arid Tropics (ICRISAT) wanted to reduce the risks associated with adoption of higher-yielding pearl millet hybrids and extend the useful economic life of these varieties, especially for poorer producers. Biotechnology helped us to achieve this. With tools from the John Innes Centre and support from the Plant Sciences Research Programme of the Department for International Development (DFID), we developed and applied molecular genetic tools for pearl millet.

We mapped the genomic regions of pearl millet that control downy mildew resistance, straw yield potential, and grain and straw yield under drought stress conditions. Then our millet breeders used conventional breeding and marker-assisted selection (MAS) to transfer several genomic regions conferring improved downy mildew resistance to the two elite inbred parental lines of popular hybrid HHB 67. We then used MAS to derive two new varieties - ICMR 01004 and ICMR 01007 - with two different downy mildew resistance gene blocks. These varieties have performed as well or better than their parent lines for grain and straw yield, and are markedly improved for downy mildew resistance.

They also retain several favourable traits, including 1 000-grain mass, panicle length, plant height and rust resistance. Hybrids based on crosses involving ICMR 01004 and ICMR 01007 have recently advanced to trials in the Indian states of Gujarat, Rajasthan and Haryana under the All India Coordinated Pearl Millet Improvement Project.

This follows their successful evaluation in 2002, in which they exhibited marginal grain yield superiority and substantially better downy mildew resistance than HHB 67, while maintaining the early maturity that contributes to its popularity. At least one of these two hybrids could be released as a replacement for HHB 67 before the latter succumbs (as it surely will) to a downy mildew epidemic.

Because HHB 67 is so widely grown by poor farmers in India, if its timely replacement could prevent such an epidemic for even one year, the losses avoided would exceed the total value of research-funding support by DFID for the development and application of the molecular genetic tool kit for pearl millet (£3.1 million to date). All future benefits from this research by ICRISAT, its DFID-supported partners in the United Kingdom, and collaborating national programme partners in India can then be considered profits to society.

THE DEVELOPMENT OF BREEDING POPULATIONS IN CROPS, LIVESTOCK, FORESTRY AND FISHERIES

The use of molecular markers to measure the extent of variation at the genetic level, within and among populations, is of value in guiding genetic conservation activities and in the development of breeding populations in crops, livestock, forestry and fisheries. Studies carried out using these technologies in fish and forest tree species have revealed high levels of genetic variation both among and within populations.

Livestock species are characterized by a high degree of genetic variation within populations, whereas crops exhibit a higher degree of variation across species. Data from other approaches, for example field observation, often cannot provide such information or are extremely difficult to collect. Molecular markers are increasingly used to study the distribution and patterns of genetic diversity. Global surveys indicate, for example, that 40 per cent of the remaining domestic livestock breeds are at risk of extinction. Most of these breeds are found only in developing countries, and there is often little knowledge about them or of their potential for improvement.

They may contain valuable genes that confer adaptation or resilience to stresses, such as heat tolerance or disease resistance, that may be of use for future generations. Modern biotechnologies can help to counteract trends of genetic erosion in all food and agriculture sectors.

PLANT GENOME ORGANIZATION AND STRUCTURE

The plant genome is organized into chromosomes that provide the structure for the genetic linkage groups and allow faithful replication, transcription and transmission of the hereditary information. Eukaryotic genomes are much more complex than prokaryotic genomes. And further, plant genomes are more complex than other eukaryotic genomes.

Prior to the development of recombinant DNA technology genomes, were analysed by reassociation kinetics techniques. Reassociation kinetic experiments are performed by melting DNA and allowing it to reanneal upon itself or with another population of either DNA or RNA molecules. The kinetics of the reassociation provide data that can be used to analyse the overall structure, evolution and expression of genomes. The most common method of denaturing duplex DNA is by heating to 100°C. But how can we monitor this denaturation and subsequent renaturation? The most common method is by measuring the absorbance change in the ultraviolet region at 260 nm.

The important property is that melted, single-stranded DNA absorbs about 40 per cent more at 260 nm than duplex DNA. If we slowly heat DNA the absorbance will increase dramatically over a short range of temperatures. The mid-point of this transition is called the melting temperature or Tm. Under physiological conditions, the Tm usually lies in the range of 85-95°C. Thus, without altering the cellular conditions, the duplex DNA is stable in the cell. The exact temperature that a particular DNA melts depends on several parameters. The GC content is important because GC base pairs have three hydrogen bonds compared to the two for AT base pairs. Thus, the higher the GC content the higher the Tm.

Table. Effect of GC Content on Tm

%GC	Melting Temperature
40	87°C
60	95°C

DNA denaturation is reversible, and the reversible process is called renaturation. This process requires that the temperature be lowered gradually.

As this lowering of temperature proceeds the following occurs:

- Single-stranded molecules randomly encounter each other.
- Short complementary stretches of duplex DNA form.
- The DNA then zips back together to form the original structure.

Obviously if a single molecule is denatured it should be able to reform completely. But if two separate molecules are denatured together, for example wheat and barley DNA, then complementary regions between the two DNAs will be able to form duplex DNA. The ability of two molecules to renature is called hybridization.

Hybridization: The pairing of complementary nucleic acids. Performing hybridization experiments in a solution is called liquid hybridization. We have already discussed the related procedure called filter hybridization.

ANALYSIS OF GENOMES BY REASSOCIATION EXPERIMENTS

If a DNA molecule is melted and allowed to reassociate, the complexity of the genome dictates the rate in which duplex DNA will form. If we consider a simple molecule that consists of alternating GCs, this molecule will be able to form a duplex quicker than a molecule that consists of repeating blocks of AGCT. As the number of different combinations of bases increases, the time required for complete duplex formation to occur will increase. Renaturation, or duplex formation requires random collisions between two single-stranded molecules.

This process follows second-order kinetics and is concentration dependent. We will not go through the derivation of the formula but the important parameter used to define a certain DNA is:Cot½. This value is defined as the amount of time required for one-half of the DNA to reanneal or form duplex DNA.

The units for this parameter is moles of nucleotides per litre per second. The more complex the genome of interest, the longer it will take for like sequences to reanneal. Consequently, the Cot½ will be larger. Thus in terms of reassociation kinetics complexity has a specific definition.

Complexity: The total length of different sequences. For example, E. coli is considered to have a complexity of 4.2 × 106 base pairs. What is the experimental procedure used to derive these values?

In general the procedure is:

- Shear the DNA to be analysed to a length of about 300 bp.
- Melt the DNA (usually in 0.12 M phosphate buffer) by boiling for 5 min.
- Quickly place at 60°C.
- Take aliquots at different time points. Separate single-stranded DNA from double-stranded DNA by hydroxyapatite. Measure the amount of DNA that is double-stranded by absorbance at 260 nm.

- Plot the amount that is single-stranded versus the Cot value. The Cot value is expressed in log equivalent. This plot depicts the Cot curve.

When this type of experiment is performed with eukaryotic DNA three components are usually seen. These components each reanneal with their own unique Cot½ value. The three components are termed the fast, intermediate, and slow components. Why do we see these three components?

Eukaryotic genomes are characterized by sequences that are represented by different copy numbers. If a sequence is found many times in the genome, it will reanneal much quicker than those sequences that are found only once in the same genome. Thus the equivalent Cot curve for a eukaryotic genome will be different than a genome, such as E. coli, which only contains single copy sequences. A comparison of the Cot value of each of these components with an E. coli standard allows us to derive the complexity of each component.

The complexity of the slow component of the genome is greater than that for the other two components and is considered to represent the single copy portion of the genome. The complexity of the slow component can be used as a good estimate of the genome size. The genome size will be the sum of the lengths of all the unique sequences. Using the example from Genes V - Lewin, p.664, the complexity of the slow component is 3×10^8 bp and the complexity of the intermediate component is 6×10^5. If we divide the complexity of the slow component into the intermediate component we get 2×10^{-3}. This demonstrates that the intermediate component contributes very little to the complexity of the genome.

Therefore, the complexity of the slow or single copy portion of the genome can be considered equal to the genome size. To derive the complexity of each component, it is necessary to run a standard, such as E. coli DNA, with each experiment.

E. coli is considered to consist of only single- copy sequences. Let's say that in the experiment from which the Cot for each component was derived, the Cot½ value for E. coli was 4. Experimentally it was determined that the slow component comprised 45 per cent of the total DNA. Therefore, if only that component was annealed, the Cot½ value would be 283 (630×0.45).

That value is 71 (283/4) times slower than for E. coli. Therefore the complexity of the slow component is 71 times that of E. coli or 3.0×10^8 ($71 \times 4.2 \times 10^6$). The complexity of the other components is derived similarly. Genome size is quite variable throughout the biological world and the genome size in plants shows the greatest variation of any kingdom in the biological world.

Table. Variation in Genome Size among Plants

Species	kb/haploid	pg/haploid
Arabidopsis	7×10^4	0.15
Lily	1×10^8	100.00

Conversion factor: 1 pg = 0.965 × 109 bp = 6.1 × 10^{11} daltons

One manner in which a genome can be described is by determining the distribution of fast, intermediate and slow components in the genome. For comparison purposes, what does the human genome look like? Distribution Sizes Among Components of the Human Genome

Component	% of Genome
Fast	6
Intermediate	38
Slow	50

The table Sequence Distribution of Selected Plant Species lists the different components from different plant species. As you can see, plant species exhibit a wide range of values for each of the components.

The genome of Arabidopsis is essentially entirely single copy sequences (the repetitive sequences have been determined to be essentially all chloroplast DNA). At the other extreme, pea and wheat genomes have only 10–20 per cent single copy sequences. Reassociation kinetic experiments of polyploid species, such as bread wheat (Triticum aestivium) were unable to derive a component that displayed true single copy kinetics. Instead the slowest component appeared to act as if it consisted of copies represented three times. This result is consistent with the current hypothesis that bread wheat was developed from the introgression of three diploid wheat species. Genomic analysis suggest that this hypothesis is correct since the slowest component appears to consist of sequences represented three times.

COMPONENTS OF REPEATED SEQUENCES

The intermediate and fast components are composed of sequences that are found many times in the genome. These sequences are called repetitive sequences and can vary in size from a 100 bp to 1000 bp or more. Furthermore, these sequences have undergone sequence divergence by the addition or deletion of sequences or by changes in the base pair sequence.

Thus, the repeated sequences themselves show some divergence. An example of a highly repetitive sequence is the repeat found to be associated with the knob heterochromatin of corn. It ranges from 3-5 × 10^5 copies on a small knob to 1 × 10^6 on the large knobs. This sequence is unique to knobs and is not found associated with any other heterochromatic regions of corn. An example of a functional repeated sequence in plants is the corn storage proteins, zeins. Two major classes of zeins exist, the 22 and 19 kd classes. Sequence analysis has shown that both classes have the same structure. The only difference between the 22 and 19 kd class is the repeat unit.

The 22 kd class has 8 repeat units and the 19 kd class has 7 repeat units. Estimates have been made of the number of copies of these genes and 30-50 copies of the 22 kd class are found in the corn genome. Thus, the repeat unit would be represented 240-400 times in the genome.

ORGANIZATION OF SINGLE-COPY SEQUENCES

Single-copy sequences are interspersed throughout the plant genome. These sequences are bounded by repeat sequences. The length of the single-copy regions varies widely among plant species.

In general, two types of arrangements are recognized:

- *Short Period Interspersion*: Single copy sequences of 300-1200 bp are interspersed as islands among short lengths of repeat sequences
- *Long Period Interspersion*: Single copy sequences of 2000-6000 bp are interspersed as islands among repeat sequences

How can the interspersion type be determined? As we discussed earlier, reassociation kinetic experiments are performed with 300 nt long fragments. These experiments give a characteristic Cot curve that defines each of the components. But what happens if the fragment that is followed is of a longer length, for example 900 nt?

This fragment could be considered to consist of three 300 nt fragments, and each fragment may be from any of the three components. Indeed this is how these experiments may be performed. A tracer of a longer length (such as 900 nt) is prepared and radiolabelled, for example with H^3. This tracer is added to a normal Cot reaction in such a low concentration that the rate is not affected.

The reaction is then allowed to proceed to a Cot value determined from a normal experiment with only 300 nt fragments where the repetitive sequences have reannealed, but the single-copy sequences have not. Then the amount of tracer which remains single-stranded is determined.

What will be seen is a reduction in the amount of expected single-copy sequences. Why? If the 900 nt long fragment contains both single copy and repetitive sequences it will reanneal under these conditions as a repetitive sequence, and thus the amount of DNA that appears in the single-copy fraction will be reduced proportionally.

Therefore, that single-copy sequence is interspersed with repetitive sequences. If for example the single copy fraction is determined to account for 40 per cent of the genome when 300 nt fragments are used, and we calculate that the genome has 30 per cent single-copy sequences when a 900 nt tracer is used, 10 per cent (40 per cent-30 per cent) of the 900 nt fragments contain single-copy and repeat fragments and that 25 per cent (10 per cent/40 per cent) of the single-copy DNA is interspersed with repetitive sequences at an interval of 300-900 nt. A final concern is the size of the repeat units. This can be obtained by using a long tracer (5000 nt, for example) and reannealing to an appropriate Cot such that only repeat sequences bind to the tracer.

The product is then treated with an enzyme that cuts only single-stranded DNA. This enzyme is S1 nuclease. After digestion you obtain products that only contain repeat sequences. These products are then sized to give the average, mode and range of repeat sequences.

EVOLUTION OF REPEATED SEQUENCES IN CEREALS

The analysis of repeated sequences has also provided experimental evidence which supports the current model of cereal speciation. The following tables and the associated diagram detail the results.

These show that once a repeat enters the lineage it remains. Therefore group I repeats are found in all of the species because it is the most ancient repeat. Secondly, new repeats are added to the lineage and appear in all species which subsequently diverged from the lineage. Finally, species-specific repeats are formed in each species after that species has diverged from the common lineage. The distribution of the lineage specific repeat supports the speciation model.

ESTIMATING THE NUMBER OF EXPRESSED GENES

Reassociation kinetics can also estimate the number and abundance of expressed genes. These experiments are performed using high concentrations of RNA and tracer levels of either DNA or cDNA. These experiments are analagous to DNA reassociation experiments except the units are expressed as moles of ribonucleotides per litre per second and the value is called a Rot.

If tracer amounts of DNA are allowed to reanneal only a portion of that DNA will form a duplex because not all the DNA will be expressed in the RNA population to which it is being hybridized. Let's use the example in the Genes V to derive an estimate for the number of genes that are being expressed. In this experiment 1.35 per cent of the DNA hybridized to the RNA.

Since RNA is single stranded the other anti-sense strand of the DNA would not have a partner with which it could hybridize. Thus actually 2.7 per cent of the DNA is represented in this RNA population. If the genome size is 8.1 X 108 bp and the single-copy sequences represent 75 per cent of the genome then we can estimate the complexity of the expressed DNA.

$0.027 \times 0.75 \times (8.1 \times 108 \text{ bp}) = 1.7 \times 107$ bp If each gene is about 2000 base pair then the number of genes that is expressed is:

$$7 \times 107 \text{ bp}/ 2000 = 8500 \text{ genes}$$

CHLOROPLAST GENOME ORGANIZATION

All angiosperms and land plants have cpDNAs which range in size from 120-160 kb; three expceptions are:

Species	Size (kb)
N. accuminati	171
Duckweed	180
Geranium	217

All cpDNA molecules are circular and spinach is used as the basis for all comparisons. Very few repeat elements are found other than short sequences of less than 100 bp. The notable exception is a large (10-76 kb) inverted repeat section, which when present, always contains the rRNA genes. (Legumes such as pea do not contain this repeat.) For the majority of species, this repeat region is 22-26 kb in size. Finally,the genetic order of the ribosomal unit is conserved in all species:

$$16S - tRNA^{ile} - tRNA^{ala} - 23S - 5S$$

Two other features of chloroplast DNA is described. First it was shown to that it can exist in in two orientations This implies that the molecule can undergo an isomerization event. Second is has been shown that spinach, corn, tomato and pea can all exist as multimers (PNAS 86:4156, June 1989).

Multimer	Relative Abundance	Per cent
Monomer	1	67.5
Dimer	1/3	22.5
Trimer	1/9	7.5
Tetramer	1/27	2.5

Because photosysnthesis is the primary function of the chloroplast it is not surprising that the chlroplast genome contains genes which encode for proteins that are involved in that process.

Reaction	Function
Dark Reactions	rbcS (nuclear encoded)
	rbcL (chloroplast encoded)
Light Reactions	apoproteins for PSI andPSII
	cytochrome b6
	cytochrome f
	6 of 9 ATPase subunits
	cab, LHC proteins (nuclear encoded)
	plastocyanin (nuclear encoded)
	ferredoxin (nuclear encoded)
Other	19/60 ribosome binding proteins
	translation factors
	RNA polymerase subunits
	tRNA and rRNA genes

Atrazine resistance is apparantley mediated through the psbA gene sequences of the 32 kd protein which is encoded by cpDNA. DNA sequence analysis revealed the following amino acid changes that are thought to be important.

Species	AA#	Susceptible	Resistant
Blue green algae	264	Ser (TCG)	Ala (GCG)
Chlamydomonas	264	Ser (TCT)	Ala (GCT)
Solanum nigrum	264	Ser (AGT)	Gly (GGT)
Amaranthus	228	Ser (AGT)	Gly (GGT)

Evolutionary Changes of cpDNA:

- The majority of changes are small insertions and deletions of 1-106bp; significantly, a few length mutations of 50-1200 bp are clusted in "hot spots".
- The largest deletion occured in pea where an entire rRNA cluster is lost.
- The most common evolutionary change is in gene order. Small changes in the gene order occur, especially in the algae, but inversions have generated large scale order changes:
  - *Legumes*: About 50 kb inversion brought rbcL closer to psbA
  - *Wheat*: About 25 kb inversion brought atpA closer to rbcL

THE CHLOROPLAST GENOME OF MITOCHONDRIAL GENOME ORGANIZATION

In comparison to the chloroplast genome, the size of the mitochondrial genome is quite variable.

Species	Size (kb)
Oenothera	195
Turnip	218
Corn	570
Muskmelon	2400

Further, in comparison to the mitochondrial genomes of other species the size is quite large and variable. For example, animal mitochondrial genomes range in size form 15-18 kb, and fungi mitochondrial genomes range form 18-78 kb. Plants may code for more proteins than with species.

For example, genes for ribosomes, subunits I and II of cytochrome oxidase and ATPase subunits are located on the mitochondrial genomes of plants. When DNA from corn mitochondria was investigated with EM, several circular molecules of different sizes were detected. Once the genome was mapped it became apparent that a mechanism existed to generated these circles of different sizes. It is now understood how these molecules arise. First, lets look at the simple situation of turnip. Two direct repeats undergo intramolecular recombination to give the two smaller molecules:

$$218 \text{ kb} \rightarrow 135 \text{ kb} + 83 \text{ kb}$$

The mitochondrial genome of corn undergoes the same type of recombination, but the events are more complex. First, the master circles can be subdivided into two major subgroups:

$$570 \text{ kb} \rightarrow 488 \text{ kb} + 82 \text{ kb}$$

$$570 \text{ kb} \rightarrow 503 \text{ kb} + 67 \text{ kb}$$

The second group of molecules are still labile and can produce several other subpopulations. Further two subgenomic circles can unite to form a larger circle. This variability is possible because corn has 10 repeats with which intramolecular recombination can occur.

Species	Master Circle Size (kb)	Sub-genomic Circle Size (kb)	Repeat Size (kb)
Turnip	218	135 + 83	2
Cauliflower	217	172 + 45	?
Black Mustard	231	135 + 96	7
White Mustard	208	none	none
Radish	242	139 + 103	10
Spinach	327	234 + 93	6

Introns have been located in the cytochrome oxidase subunit II gene (the cytochrome complex consists of 3 mitochondrial and 4 nuclear encoded genes).

This gene contains one intron in rye, corn, wheat, rice, and carrot, but for other species such as Oenothera, broad bean, cucumber the gene has no intron.

PROMISCUOUS DNA

Stern and Lonsdale (1982) hybridized mtRNA to a SstII digest of maize mt DNA and found that it hybridized to fragments known not to contain mt rRNA genes. The question of interest was - what was it hybridizing to? They next looked at a cosmid clone of corn mtDNA that hybridized to the mt RNA and found that it hybridized to a RNA molecule of the size of the cp 16S RNA gene.

How could this have happened? They next mapped the clone and compared it to the map of the corn cpDNA and found that the clone map was almost congruent with that of the the cpDNA 16S RNA region. This mapping showed that the two maps were nearly identical over a 12 kb region of DNA. These results suggest that cpDNA had been transferred to the mitochondrial genome. The observation that organelle DNA was found in other DNA compartments of the cell was extended by other researcher.

Stern and Palmer looked at corn, mung bean, spinach and pea and found extensive evidence of cpDNA/mtDNA homology. These observations were extended to other DNA locations in the plant cell. Kemble *et al* (1983) demonstrated that mitochondrial DNA sequences are located in the nucleus of corn. Scott and Timmis (1984) showed that cpDNA sequences are found in the nuclear DNA.

RNA EDITING

The Central Dogma of Molecular Genetics states that the information that is found in DNA is used to produce mRNA molecules that are instrumental in the production of proteins.

Therefore, the information flows directly from DNA to protein, via the RNA intermediate molecule. Recently it has been discovered that the information that is contained in the DNA is not always found in the RNA products used to make proteins. It has now been demonstrated that mitochondria and chloroplast contain the biochemcial machinery to alter the sequence of the final transcription product. This process is called RNA editing. This process was identified in the following manner. Sequence analysis of a number of cytochrome c oxidase subunit II genes from non-plant species revealed that a tryptophan residue was invariant at several locations in the final protein product. But sequence analysis of this gene in several plant species revealed arginine at those positions.

Since a single base pair change in the codons for the two amino acids could generate this change (CGG for UGG), it was suggested that CGG encoded for tryptophan and not arginine in plant mitochondria. (This is the only change in codon usage that has been suggested for plants and has been postulated for several other genes as well.)

But this change in codon usage was not universal, that is some CGG codons actually specified arginine in the final protein product. Furthermore, no amino acyl tRNA that recognized CGG was found to be charged with tryptophan, a prerequisite if this codon specification was actually real. The solution to this dilemma was found by sequencing the mRNA products for cytochrome oxidase subunit II genes. It was found that in the mRNA the cytosine residue had been changed (edited) to uridine at the sequence location where the invariant tryptophan residue is found.

This changed the codon at that location to UGG which is recognized by a tRNA that carries the amino acid tryptophan. An analysis of three other plant mitochondrial genes where the same altered codon usage was predicted suggested that mRNA editing was also occurring at the codon and that a cytosine residue was edited to uridine. This editing process has also been detected in protozoa and it remains to be determined if RNA editing is a widespread function in mitochondria. A final point that this editing function highlights is that the sequence that is found in the DNA is not entirely and faithfully represented in the final protein product.

4

Cloning of Genes

VARIETY OF CLONING AND EXPRESSION

A variety of cloning and expression vectors, which are used for a variety of experiments involving cloning of DNA segments or genes: This is achieved through construction of chimeric DNA molecules for the purpose of obtaining molecular probes and gene libraries.

These different uses of vectors and the methods of cloning will be briefly described. A brief account on restriction enzymes is also included, since these enzymes are used regularly as a routine in all DNA cloning and recombinant DNA experiments.

DNA CLONING OF RESTRICTION ENZYMES

For cloning of DNA, often we need to cut the DNA at specific sites, which are recognized and cleaved by specific - enzymes (endonucleases, which cleave DNA at internal sites), described as restriction enzymes. Also by locating the positions of cleavage sites of a number of restriction enzymes, restriction maps can be prepared. These restriction enzymes recognize short sequences of double stranded DNA as targets for cleavage. Different enzymes recognize different, but specific sequences, each ranging in length from 4 to 8 base pairs. Each enzyme is named by a three letter (or four letter) abbreviation (letters are italicized) that identifies its origin.

Roman numerals (I, II, III, etc.) are added to distinguish several enzymes with same origin. For instance EcoRI is derived from E. coli and Hpal is derived from Haemophilus parainfluenzae.

Besides cleavage, modification in the form of methylation is also brought about by some enzymes called modification enzymes (sometimes also called methylases). This methylation distinguishes genes in different states of functioning. There are also enzymes which perform the function of both restriction (cleavage) and modification (methylation). Based on these attributes, restriction enzymes have been grouped into two classes:

- Types II restriction enzyme systems (*e.g.* EcoRI) which have separate enzymes for restriction and modification.

- Type I (EcoK and ECoB) and type III (EcoPl, EcoP15) enzyme systems, in which same enzyme possesses both activities (bifunctional), although the restriction and modification sites differ in positions.

Of the above two classes of restriction enzymes, type II enzymes are most important for cloning purposes. The target sites for these enzymes are 4-8 bp with a symmetry (palindromes).Most type II restriction enzymes cleave DNA at unmethylated target sites.

Some enzymes introduce staggered cuts, others generate blunt ends. Enzymes with 4 bp target sites are used when frequent cuts are desired and those with 8 bp (eg. Not) are used when rare cuts are desired to get long DNA segments. Otherwise majority of enzymes used have 6 bp target sites. Some of them can cleave both methylated as well as unmethylated targets, but majority of them cleave only unmethylated targets.Recognition sequences of several restriction endonucleases (an arrow in each case indicates the position of cleavage)

RECOMBINATION FREQUENCIES IN TECHNIQUE OF RESTRICTION MAPPING

Genetic maps are prepared using recombination frequencies. However, the recombination frequencies, which are the function of distances between genes, are not completely independent of:

- The nature' of mutants used.
- The position of these mutants on chromosomes.
- The genetic background.
- The environmental conditions.
- A variety of other factors.

Consequently, the distances between genes on a genetic map may not correspond to the distances between them on the DNA molecule of which they are a part at the molecular level. Gaps may also be present on a genetic map due to non-availability of mutants in that region.At the molecular level the fine structure of a gene can be studied through determination of nucleotide sequence of the concerned DNA segment. Sometimes this is done through isolation of a DNA segment corresponding to a gene followed by sequencing of the DNA segment, which needs time and energy. Instead we can prepare a map of the DNA by its cleavage at specific sites with the help of restriction endonucleases, which recognize very short specific DNA sequences and cleave the DNA at these specific sites.

These sites of cleavage can be identified and mapped to give rise to a restriction map. A restriction map thus consists of a linear sequence of sites, each' tor a specific enzyme and the distances between them are measured in terms of number of base pairs of DNA. This technique can be used both in prokaryotes and eukaryotes, although all regions of chromosomes can not be easily mapped in eukaryotes.

RESTRICTION CLEAVAGE AND GEL ELECTROPHORESIS

If we take a particular DNA molecule or DNA sample, digest it with a specific restriction enzyme and then subject the sample to gel electrophoresis (a technique in which DNA digest is loaded on gel slab and the DNA fragments move under the influenoe of an electric current), we will notice a series of bands on the gel slab or cylinder.

Agarpose Gel Electrophoresis: The bands can be observed under ultraviolet light after staining with ethidium bromide. The position of different bands will depend on DNA fragment size, such that smaller the fragment, more rapidly will it move, and longer the fragment, more slowly will it move. It will mean that the fragment away form the loading site will be smaller and those close to the loading site will represent longer DNA fragments.

The gel can be calibrated by using a mixture of DNA fragments of known lengths, so that the positions of the bands of known lengths on this standard gel can be compared with the positions of bands in the experimental DNA digest. The fragment length for each band of DNA digest can thus be determined sometimes using a computer device. we have shown the results of digestion of 5000 bp long hypothetical DNA molecule digested separately by two enzymes A and B.

The enzyme A cleaves the DNA into four fragments of lengths 2100,]400, 1000 and 500bp, while the enzyme B cleaves it into three fragments of lengths. 2500, 1300, 1200bp. These data with some additional experiments, can be used to generate a restriction map as shown in the next section. Technique of double digests to determine cleavage positions of DNA due to one wnzyme with respect to those due to another enzyme (the two enzymes are A and B);

DNA DIGESTION

The data on DNA digestion by more than one endonucleases as discussed above can be utilized to arrange the sites of cleavage in a defined order. This can be done by several methods and one of them is illustrated. This method involves successive digests with two individual enzymes, where we extract each fragment produced in the individual digests with either enzyme A or enzyme B and then cleave it with the other enzyme.

The original DNA sample is also digested by a mixture of both the enzymes to confirm the results of individual successive digests. We have shown results of such an exercise.

When the fragment A-21O0 is obtained and digested with enzyme B, it is cut into two fragments of 1900 bp and 200 bp. Fragment B-2500 (obtained from individual digest with enzyme B) is similarly cut by enzyme A into two fragments of 1900 bp and 600 bp, suggesting of A-2100 and B- 2500, in the region of fragment of 1900 bp which is obtained by one cut with enzymes A and B (double digest). It may be that in reciprocal digests (A followed by B and B followed by A) and double digests, same fragments are available. Using this information, we can field the overlapping regions in A and B degests and find out the sites of cleavage by A and B.

HYBRIDIZATION IN RESTRICTION MAPPING

The above technique of individual and double digests can be supplemented with other techniques for actual construction of maps. For instance, by permitting incomplete digestion, fragments longer than those obtained by complete digestion 'may be obtained. These will be called partial digests.

In the example used in the last section, in partial digests with enzyme A, instead of getting A-I 000, A-2100, A-1400, A-500, we may,get A-3100, A-1400 and A-500, which will suggest that A-l000 and A-21oo lie in adjacent regions. In another partial digest if we get A-l000, A-3500 and A-500, this will suggest that A-2l00 and A-1400 are adjoining, which will mean that. A-l000 and A-14oo are found on the two sides of A-2100.

This technique will thus allow the arrangement of fragments in' a linear order. Another useful technique involves labelling of the ends (end labelling) of DNA molecule before digestion, so that the fragments, containing these ends can be identified due to labelling, even after digestion. In the above example, using end labelling, if A-l000 and A-500 are found to be radioactively, labelled, these will be present at the two ends of the restriction on map.

Some of the features of a restriction map can also be confirmed by nucleic acid hybridization. For instance, in the above example,if A-2l00 and B-2500 have an overlapping region of 1900bp, they should hybridize with each other, which will confirm overlapping. The above technique will help in accurate completion of a restriction map, but will require that we have a complete set of restriction fragments which make the entire DNA region being mapped.

Once the restriction map (which is a physical map) is ready, it can be compared with the genetic map. Although large changes, already located on the genetic map, can be easily located on the restriction map, but point mutations can not always be easily located on the restriction map, since the restriction sites often do not change due to mutation.

In such a situation, one may like to determine the nucleotide sequences of individual fragments and compare them in normal and mutant individuals. Since sequencing of nucleotides is more laborious and can not be easily undertaken for the whole genome, another technique of using a variety of molecular genetic markers (*e.g.*, RFLPs, RAPDs, STs, STMS, AFLP, etc.) has been utilized during the 1980s and I 990s to generate genetic linkage maps in human, mouse, rat, pig, sheep, fruitfly (Drosophila), nematode (Coenorhabditis elegans), yeast, maize, wheat, rice, barley, tomato, potato, lettuce, rice, pepper, pea, etc.

CHIMERIC DNA MOLECULES

Chimeric DNA molecules are producted by inserting a foreign DNA segment into the DNA molecule of vector. These chimeric DNA molecules are the means of cloning genes or other small DNA sequences. The construction of chimeric DNA, as we know now, involves joining of the broken ends of the vector with the two ends of the sequence to be cloned.

There are several methods available for the construction of this chimeric DNA.

CLONING DNA FRAGMENTS IN PLASMID VECTORS

Cloning DNA fragments with homologous cohesive ends. This is the most common method, where due to the presence of a palindromic sequence, a restriction enzyme causes staggered cuts producing short complementary sequences at the two ends of (a palindromic sequence is one, which has complementary sequences at the two ends of a single strand, *e.g.* ATC GAT). This can be illustrated using the example of enzyme EcoRI, which cuts the sequences at specific positions.

When another DNA molecule is similarly cut by the same enzyme, similar sticky ends having same sequences in the single stranded ends will be produced, so that when it is mixed with the previous molecule similarly treated, the two will anneal producing a chimeric DNA or chimeric plasmid, if plasmid DNA is involved

Double Stranded DNA Fragement: Enzyme DNA ligase helps in joining the bonds at the cut ends of the two molecules. This method of cloning is commonly employed with pBR322 and pUC vectors. The above technique has the advantage of regenerating two restriction sites (*e.g.* EcoRI sites) in the chimeric DNA, so that the foreign DNA segment can be retrieved rather easily from the cloned copies of chimeric DNA by cleavage again with the same enzyme.

There are also following disadvantages with this technique. (i) The cleaved ends of two vector molecules or those of two foreign DNA segments may join end to end before the foreign DNA gets inserted.Therefore, while isolating chimeric DNA, one will have to select chimeric molecules each having only a single insert. This can be achieved by separating molecules of different sizes by gel electrophoresis.

Vector DNA and Foreign DNA to be Cloned: This difficulty can be overcome, firstly, by adjusting the concentrations of two types of DNA and *secondly,* by removing 5' phosphate groups in the linearized vector. The recognition site, particularly in the sequence to be cloned may not lie at a convenient position. so that sometimes only a part of the desired segment will be inserted.

DNA FRAGMENTS WITH HETEROLOGUS PROTRUDING TERMINI

This is the most efficient method for cloning, and any other method should be tried only when this method can not be used. In this method, the fragments to be cloned are generated by digestion with two different restriction enzymes. Similar heterologous termini are generated in the plasmid vector by removing a small fragment due to cleavage with two different enzymes.

In this method. since the two termini are not compatible. Self ligation of either the linearized vector or the foreign DNA fragment leading to circularization is eliminated. Moreover, chimaeric DNA molecules will contain each only a single copy of the insert DNA oriented in a defined direction.

Recombinant molecules with two or three copies of insert will be very rare (this is unlike the case with cloning of DNA fragments with homologous cohesive ends). As an illustration of directional cloning, pUCl9 can be digested with BamHI and HindIIIand the resulting large fragment can be purified, removing the small fragment. The vector can then be ligated (using T4 ligase) to a segment of foreign DNA that contains termini compatible with those generated in the vector due to BamH I and Hind III. The steps involved in this directional cloning. In this method, DNA can be cut at the desired position both in the vector and in the clone, without staggered cleavage.

Using precursor dA TP, poly dA is added at both the 3' cut ends of the vector with the help of enzyme terminal transferase. Similarly using precursor dTTP poly dT is added at both the 3' cut ends of the DNA sequence to be cloned.

The vector and clone can then be joined by annealing the poly dA with poly dT tails and then ligating them using DNA ligase enzyme In this techinque, there is no chance of reannealing between the two cut ends of the same DNA molecule or of the two similar DNA molecules, so that one of the disadvantages of the first method has been overcome in this case. However, in this technique it will not be easy to retrieve the cloned DNA, because the recognition site of the enzyme has been lost in this case due to insertion of poly dA and poly dT. But if poly dG and poly dC are used instead (which serve the same purpose), cloned sequence can be retrieved, since poly dG and poly dC regenerate the recognition site for the enzyme PstI.

BLUNT END LIGATION BY T4 DNA LIGASE

In this technique, a restriction enzyme is used to cut a duplex DNA at the same place in both the DNA strands. The broken ends are then used for joining with the two ends of another DNA molecule irrespective of the sequences present at the broken ends of the two DNA molecules. The T4 DNA ligase is used for this joining reaction. The disadvantage of this technique is that any two broken ends may join including those belonging to the same DNA molecule. This leads to the production of a variety of products and one will have to select the desired product from a mixture of products. The blunt end ligation is.used for developing a method in which the DNA to be cloned can be easily retrieved whenever required. This method makes use of short DNA duplexes (linkers), that contain EcoRI palindrome or some equivalent palindrome, which being small in size, can be synthesized chemically.

These linkers can be linked to the blunt ends of vector DNA by blunt end ligation. This will allow the creation of an EcoRI site in the linker region of the vector, so that when a DNA segment is now cloned, the inserted DNA segment can be retrieved by cleavage with EcoRI.

CLONING IN PHAGE Λ AND COSMID VECTORS

Cloning in phage l. vectors. Phage λ vectors are used for a variety of purposes including the following:

(i) subcloning of DNA fragments that are propagated in vectors with large capacity (*e.g.* cosmids)

(ii) construction of genomic and cDNA libraries.

Cloning in phage λ vectors involves the following steps:

- Digestion of phage λ with a restriction enzyme (digestion with two enzymes, one followed by the other is sometimes carried out to remove the stuffer; this is necessary in case of only some of the replacement vectors) and purification of phage l; purified arms of several popular A vectors including those for EMBL3 and EMBL4, and those for ëgt10 and ëgt11 are also available commercially, so that if the arms are procured commercially, this step involving digestion and purification can be dispensed with;
- Ligation of A arms to fragments of foreign DNA to be cloned;
- Packaging of recombinant phage A in vitro, utilizing the packaging - extract that is available commercially or can be prepared in the laboratory (packaging extract is a lysate of phage A infected E. coli cells and contains empty phage heads, unattached phage tails and phage-encoded proteins needed for DNA packing); this is followed by infection of E.coli competent cells by packaged recombinant. A vectors, although rarely transfection with naked A DNA without packaging can also be tried;
- Multiplication of transformed E. coli cells and (v) lysis of E. coli cells to release recombinant phage, forming plaques and recovery of cloned DNA segment.

TYPES OF GENETIC MARKERS

Up to the mid-1960s the only easily usable genetic markers were those that produced clearly visible effects on the plant's phenotype such as colour, size, shape, disease resistance, etc.. All these were the result of mutations to alleles which had, obviously, a major effect on the development of the plant. Such mutants were rare and generally not desirable in commercial material although they may be selected by breeders of decorative plants for their novelty value.

As genetic markers, however, they have many disadvantages. They often affect the fitness of the individual and could well have effects on important agronomic traits, *i.e.* they exhibit pleiotropy. Because they are rare, it would be necessary to bring several such mutants together in the same material in order to use them, and they would not normally be present in commercial cultivars or, indeed, in wild material. This not only takes time, but the combined effects of many such alleles would mitigate against their value as a genetic tool. In the 1960s through to the 1980s, naturally occurring genetic polymorphisms at the protein level came to be easily recognized. Variation in enzymes and storage proteins was detectable on an electrophoretic gel for a very high proportion of gene loci which could be studied in this way and, unlike major mutants, such naturally occurring allelic variants had much smaller effects on fitness and hence were more common and existed naturally in populations. The number of such polymorphisms was, however, insufficient to provide the coverage of the genome desired by breeders and conservationists. The development of techniques during the late 1980s up to the present revealed the vast amounts of genetic polymorphism which exists at the DNA level and has revolutionized genetic analysis, opening up a whole new range of genetic tools.

As described elsewhere in this book, this variation arises through the existence of occasional base changes in the DNA which can be recognized by restriction enzymes or by primers used in analysis involving the polymerase chain reaction (PCR). Much of the DNA, possibly as much as 90 per cent in many species, is non-coding and hence reliance on morphological or biochemical variants was only using the restricted variation that plants could tolerate within the coding parts of their DNA.

Clearly, there are limits to the tolerance of such variation. Variation in the non-coding regions, whether in the intergenic regions or within introns, is probably under far less constraint by natural selection, and hence the number of polymorphic sites is potentially enormous. Because most of these sites are outside coding genes they are referred to as loci rather than genes; genes are coding regions whilst most marker loci are not. Depending on how the DNA polymorphism is studied, various types of marker are used. Restriction fragment length polymorphisms (RFLPs) have been the most widely used until now. They have the advantage of being codominant and hence all genotypes in a cross can be identified. On the other hand they require quite large quantities of DNA, and hence plant material, and generally rely on radioactive probes, although fluorescent techniques are becoming more popular.

As a result, RFLPs are expensive to use. The use of radioactive probes incurs safety considerations, as well as taking longer to visualize the bands. PCRbased markers are now being used more widely.

Randomly amplified polymorphic DNAs (RAPDs), for example, require far less material because only small amounts of DNA are needed and the required sequences are then amplified. They do not involve radioactive probes and are both quick and cheap. On the other hand RAPDs are dominant although it is possible to distinguish heterozygotes from homozygotes by the amount of product. Amplified fragment length polymorphisms (AFLPs) offer even cheaper and more easily identified and extensive PCRbased polymorphism. Because of the number of polymorphisms which exist on individual gels and the speed of production, computer software is necessary to scan the banding patterns on the gels and to analyse and assimilate the information that is generated. Other popular markers are microsatellites (short sequence repeats) and cleaved amplified polymorphisms (CAP)-based primers. All the genetic variants discussed above will be subsumed under the broad title of marker loci. They may be of little interest in their own right, and this is particularly true of the molecular polymorphisms; their value lies in their use as markers of more useful genes.

GENE MAPPING

An important, though by no means an essential, step in genetic analysis is to produce genetic maps of the marker loci. Such maps are often referred to as 'framework maps' because they provide a framework within which important genes can be located, as well as providing a means of comparing chromosome organization in other closely or distantly related species. There are two stages to mapping.

Firstly, to arrange the markers in a linear sequence separated by an appropriate map distance, *i.e.* to construct a linkage map. The second is to relate the linkage maps to particular recognizable chromosomes. The latter is often the most difficult and generally not essential for breeding or conservation work. We will, therefore, concentrate on the former. Chromosomes contain a single linear molecule of DNA and hence the markers on that chromosome occur at particular positions along that molecule.

Typically, each chromosome contains 107 to 108 base pairs (bp), *i.e.* 104 to 105 kilo-base pairs (kbp) of DNA, while a typical structural gene, coding for a polypeptide chain, would be between 1 and 2 kbp long. Assuming only 10 per cent of the genome is coding DNA, then a chromosome probably contains something of the order of 1000 to 10 000 genes. Fortunately, the fact that the chromosome is a linear molecule means that the genes need to be mapped in only one dimension.

Currently, the only useful method of gene mapping, at least as far as breeders and conservationists are concerned, relies on recombination between homologous chromosomes resulting from genetic exchange, which can be seen as chiasmata at diplotene through first metaphase of meiosis. A single chiasma on a particular chromosome results in half the gametes from such a meiosis being recombinant for that chromosome: half the gametes will contain a copy of that chromosome that is part maternal and part paternal in origin, while the remaining gametes will be entirely parental. Very few plant species have a sufficiently low number of clearly recognizable chromosomes that the number of chiasmata on a particular chromosome can be compared in different nuclei.

However, in many species the total number of chiasmata in each diplotene nucleus can be counted, at least in pollen meiosis, and the average number of chiasmata per nucleus calculated. The number and position of chiasmata on a particular pair of homologous chromosomes will vary from nucleus to nucleus; it is as though there are a large number of potential sites of exchange, but that in any given meiosis only a few sites are actually involved. The longer the chromosome the more potential sites there might be. There will invariably be one chiasma and there may be two, three, four or more; a typical set of results is shown in Table A chromosome that has one chiasma on average is said to be 50 centimorgans (cM) long – a map length unit named after the American geneticist Thomas Hunt Morgan.

This relates to the fact mentioned above that one chiasma results in 50 per cent recombinant chromosomes. By extension, a chromosome with an average of 2.5 chiasmata is said to be 125 cM (*i.e.* 2.5 350 cM) long. It appears that, as a rough rule of thumb, each chromosome has on average two chiasmata and so is 100 cM long. It follows, therefore, that providing the haploid chromosome number is known (n), the total map length will be approximately 2 3 n 3 50 cM. The average number of chiasmata per nucleus has been calculated in many species and this rule generally holds.

Table. Chiasma Fraquency and Mapping.Chiasma Frequency in Chromosomes of Secale Cereale.

Number of chiasmata per chromosome	Frequency	Percentage unrecombined Chromosomes
1	0.267	50.0
2	0.716	25.0
^{3}3	0.017	12.5
Mean	–	22.0

This rough rule is important because it gives the geneticist an idea of the total genetic map length that should be expected as more genes or markers are mapped. It provides a guide to the extent to which the currently mapped markers cover the full genetic map. Because the map is based in chiasma units, it is not the same as a physical map. The distribution of chiasmata is not uniform over a given chromosome in a species and will vary with the chromosome and the species.

Although a knowledge of the number of chromosomes and, ideally, the mean chiasma frequency per nucleus for the species, indicates the total map length, the actual map has to be constructed from examining the frequencies of progeny in crosses; *i.e.* maps are constructed from information based on the consequences of chiasmata, namely recombination of genetic markers. Recombination of genetic markers is identified by the presence of gametes that contain recombined markers, and such gametes can be recognized from the phenotypes of progeny of a cross.

The frequency of such gametes is an estimate of the recombination frequency between the two markers concerned. The closer two markers are to each other on the chromosome, the less likely a chiasma will occur between them and the less likely they are to recombine. Providing the two markers are sufficiently close that either none or one chiasma occurs between them but never more, then the recombination frequency as a percentage is equal to the map distance in centimorgans as described above. The problem arises when the markers are sufficiently far apart for two or more chiasmata to occur between them in some meioses. It can be shown that not only does a single chiasma between a pair of markers result in 50 per cent recombination between them, but so also do two, three or more chiasmata, on average. It is necessary to say 'on average', because there are a variety of possible consequences with two, three or more chiasmata in any given meiosis but, in practice, when one is looking at the progeny of a cross or self, each progeny will be the result of gametes from different meioses. This fact implies that although map distance increases linearly with the number of chiasmata, the same is not true with the frequency of recombination, which can never be more than 50 per cent for a pair of markers even though they might be 100 cM apart at opposite ends of a chromosome.

These relationships are illustrated in Figure. In order to overcome this problem, mapping functions have been devised to correct for multiple chiasmata. The most common of these are the Haldane (1919) and Kosambi (1943) mapping functions. Haldane's assumes that the probability of none, one, two or more chiasmata in a given interval follows a Poisson distribution, *i.e.* that the chiasmata are independent, random events. Kosambi's method allows for a certain degree of non-independence in chiasma occurrences. It has long been known by cytogeneticists that chiasma interference occurs over quite large regions of a chromosome. Interference means that if a chiasma occurs at a particular point on the chromosome, the next one will never occur closer than some fixed distance, the interference distance, from it. Beyond this distance, there is a short distance in which the interference disappears and the next chiasma can then occur randomly outside this range. Observational data in various species of plants are confirming this and suggest that the interference distance is generally between 15 and 20 cM, *i.e.* 15 to 20 per cent of a typical chromosome on either side of the chiasma. This has important and useful consequences for genetic and breeding work as will be shown later.

Recombination frequencies between pairs of markers can be scored in a wide variety of different crosses. The simplest to use are generations derived from an F_1 because only two alleles are segregating, and their distribution in the chromosomes of the parents of the F1 can be determined. The generations that can be used are F_2, backcrosses (Bc), recombinant inbred lines (RILs) or doubled haploid (DH) lines, with F2s being the most informative.

General formulae for calculating recombination frequencies are given by Allard. In outbreeding species where F1s are not available, a given individual can be considered as an F1 and its selfed progeny as being an F_2, even though the inbred parents do not exist. If it cannot be selfed but controlled crossing to another individual is possible, then again those genes segregating in the cross can be mapped although the situation is more complex.

There could be from two to four alleles segregating at each locus and, with respect to any one locus, the cross could represent an F2, Bc1 or Bc2. Moreover, the distribution of alleles in the two parents cannot be known and has to be inferred from the progeny. This complexity is compounded by the large number of marker loci that may be segregating in any given cross. Fortunately, a range of software packages are available to estimate these recombination frequencies, identify linkage groups, assign the markers to the most likely order and space them in map units (cM) on these linkage groups; examples are MAPMAKER and JOINMAP. Ideally the number of linkage groups should be equal to the chromosome number in the gametes but, unless the markers available provide a good coverage of the genome, it is likely that those markers on a given chromosome may appear as two or more separate linkage groups simply because the subsets are not sufficiently close for them to be recognized as being together on one chromosome.

A typical marker framework map is shown in Figure for the chromosomes of *Brassica oleracea.* Clearly these maps provide a very detailed coverage of the chromosome. Many of the markers are very close, *i.e.* less than 10 cM, and over this distance it matters little which of the two mapping functions is used. With the more widely spaced markers, *i.e.* recombination frequency >15 per cent, Haldane's function will exaggerate their distance apart because it will allow for more double crossovers than actually occur.

As more markers are mapped, the total length should converge on the value predicted by the chiasma frequency, *i.e.* approximately 2 3 *n* 3 50 cM. It is important to remember that genetic map distances have standard errors which are dependent on the size and type of population used to construct the map, the mapping population. The map obtained is that which best fits, given the data and the assumptions underlying it, but there may well be other maps which also fit the data well, though are slightly less likely. Any particular map always develops a greater aura of respectability once it is published, frequently without stating its reliability, and subsequent yet different maps are treated with suspicion. A recombination frequency of true value *p* has a standard error of ?[*p*(1 2 *p*)/*N*], where *N* is the size of the gamete population. For example, if *p* is 0.1 (*i.e.* RF = 10 per cent) and*N* = 100, the standard error is 0.03 or 3 per cent. In other words the 95 per cent confidence interval of the estimate lies approximately between 4 per cent and 16 per cent. There are other reasons, apart from statistical sampling, why maps could be wrong. The most obvious is the accuracy with which the data are scored and recorded. Autoradiographs and banding patterns on gels can easily be misread and research workers are loath to discard data even though its interpretation is ambiguous. All data should be checked independently by two or more people and any ambiguities removed.

Wrong scores will bias the data and often exaggerate map lengths as they suggest double recombination; wherever double recombination appears to have occurred in two short, adjacent intervals, the data should be checked. Changes in methylation rather than scoring errors could cause a restriction site to appear or disappear so giving the false impression of double crossing- over, or it could simply be an error.

Where maps appear to have major inconsistencies in different crosses, different chromosomal structural arrangements may be responsible, such as translocations or inversions. There is considerable evidence that chiasma frequencies, and hence recombination frequencies, may be quite different in male and female meioses even on the same plant although in other studies this may not be so.

Thus although the order of markers on a linkage group may remain the same, the relative distances between them may be quite different if the map derives from male versus female meioses. For example, in *Brassica* it appears that the genetic map obtained from a backcross where the F1 is the female parent is 60 per cent longer than when the F1 is the male parent. It is also clear that environmental factors during meiosis, particularly temperature, can affect the number and distribution of chiasmata, and this could be very critical if the crosses are set up at different times of the year.

Two or more different populations will almost certainly be segregating for different combinations of polymorphic markers, although some at least should be in common. These common ones can be used to overlay the two or more maps and a consensus map produced from an amalgamation of these. Again software is available to produce these consensus maps. Despite the various error-causing factors discussed above a considerable degree of consensus can be produced from such populations and, considerable similarity of chromosome sequence is conserved between even distantly related species, allowing the potential for cross-species genetic transfer.

It was stated above that the total map length should be dictated by the chiasma frequency, in so far as this can be accurately measured. When the first few genes were initially put onto genetic maps in the first 40 years of this century, they only covered a small part of the genome – except in wellstudied species such as *Drosophila* and maize. As more and more genes were placed on the map, so the total map lengths increased towards the asymptote dictated by the chiasma frequency. However, during the first decade following the use of molecular markers, the map lengths of some species appeared to exceed that expected, and it was even argued by some that the chiasma theory of recombination might be wrong.

No one would wish to argue that chiasma frequency data are free from error but there would have had to have been quite excessive underscoring to result in the disparity which was occasionally found. It would appear, however, that the excessive lengths were due in part to errors in scoring and to the use of small mapping populations, because subsequent map sizes have shown a progressive decrease towards the predicted asymptote. Scoring errors bias estimates of RF generally upwards, while RFs have to be large to be detectable in small populations. Our own experience with mapping in cereals and brassicas has also shown a progressive decrease in estimated map length as more markers can be found to fill some of the large gaps in the map and as errors are removed from the mapping data.

LOCATING GENES OF MAJOR EFFECT

The use of extensive molecular and other markers as described above provides a general framework map. Although this has value in its own right for comparing linkage groups in different species as well as possible structural variation within a species, the main value lies in providing a set of markers to locate genes of economic or special scientific interest. Locating individual major genes is a straightforward development of the procedures above. Let us assume that a cultivar is identified which contains an allele of interest which changes the phenotype in a clearly recognizable fashion and which appears to segregate as a single gene in crosses to other cultivars. Using a knowledge of the positions of existing marker loci, a small subset of markers can be chosen which provide a reasonably even coverage of the genome, say every 20 to 30 cM, *i.e.* approximately five per chromosome. These then have to be shown to differ between the two cultivars and any which are monomorphic replaced by polymorphic markers.

A rough position of the gene of interest can then be obtained by bulked segregant analysis. This involves taking the mapping population, *e.g.* a Bc, and dividing the plants into two groups depending on whether or not they show the phenotype of interest. Bulk samples of DNA are taken from each group, and the two samples assayed for the marker systems chosen as in the paragraph above. Any marker which is unlinked to the gene to be located will produce similar banding patterns in both samples because it will be segregating independently of the target gene. Conversely, should one of the markers be very closely linked to the target gene, it will co-segregate and the two samples will show quite different patterns for that marker.

In practice, complete co-segregation with any one marker is not very likely but, given a reasonable spread of markers over the genome, one or two are likely to show a strong association with the two groups. Thus although bands associated with both marker alleles will be found in each sample, their relative intensities will be quite different because of the association with the target gene. Having located the chromosome and the region on the chromosome, the actual position can be identified in a segregating population using the principal marker associated in the bulked segregant analysis together with other polymorphic markers situated about 10 cM on either side of the principal marker.

The target gene is then mapped with respect to the three markers. This approach can be used for a wide variety of populations and traits. Even if the trait has poor penetrance, the affected group will clearly show the marker banding pattern even if the non-affected group consists of a mixture due to misclassification.

MAPPING GENES CONTROLLING QUANTITATIVE TRAITS

Quantitative traits such as yield, quality, height and flowering time, which are controlled by several genes and are greatly influenced by the environment, create particular difficulties for gene mapping. The problems arise because the genotype for the trait concerned can never be clearly identified from the phenotype; many different genotypes could produce the same or very similar phenotypes whilst the same genotype can result in different phenotypes depending on what may be elusive factors in the environment.

Whereas with a single gene difference controlling a major effect, two or more Mendelian phenotypic classes might be recognized in ratios of 3:1, etc., quantitative traits resulting from the joint segregation of many genes show a continuous, often normally distributed range of phenotypes. The total phenotypic variation in an F_2, for example, VP, is made up of genetic and environmental components, VG and VE. The genes underlying such traits are variously called polygenes, effective factors or, more recently, quantitative trait loci or QTL.

However, it is still possible to map and measure the effects of the QTL by techniques analogous to those used for single major genes, *i.e.* by looking for the effects of co-segregation of particular marker loci with differences in the quantitative trait. Various mapping populations can be used as before, F_2, Bc, RIL, DH or open-pollinated full sib (FS) or half sib (HS) families, and the individuals or families are scored both for their phenotype for the quantitative trait(s) and for their genotype at the marker loci. The type of population that is used will depend on the traits studied and the breeding system of the plant species but, other things being equal, an F_2 population will normally provide the most information for a given size.

Unfortunately, most traits of economic importance such as yield, quality and disease resistance are not usefully scored in an F_2 population because one is interested in how the material performs in something similar to the high-density monoculture that it will encounter in agricultural practice. Measuring yield in an F_2 population with spaced plants is straightforward and statistically very powerful but it may well not provide any insight into how yield is controlled in agricultural practice.

It is therefore necessary to use a plot structure for most traits in an attempt to simulate commercial conditions. For this reason most plant breeders will prefer to work with genotypes that can be extensively and easily replicated such as RILs, DHs or F_3s. It is also easier to explain the principles underlying QTL mapping using a population of DH lines, and so for both reasons the methodology will be illustrated with DH lines, although the same principles apply to all populations. Because several trait loci are concerned, probably on different chromosomes, bulk segregant analysis is of limited applicability, at least initially.

THEORY OF QTL MAPPING IN DOUBLED HAPLOID LINES

Doubled haploid (DH) lines can be produced parthenogenetically from an F_1 by microspore or ovule culture and, in some species, such as wheat and barley, by wide outcrossing. Each original DH plant is derived from a single gamete, which subsequently becomes homozygous and disomic either naturally or by treatment with the drug colchicine, which inhibits spindle formation, and hence chromosome disjunction, at metaphase of mitosis.

Selfed seeds from such a plant produce a DH line of identical individuals so the line becomes effectively immortal. If we consider an F1 that is heterozygous for a marker locus M1/M2 and for a linked QTL locus Q1/Q2, where the 1/2 indicates the allele with an increasing or decreasing effect on the trait, this will produce the gametes and hence the DH lines shown in Table. If the DH lines are scored for some trait for which the Q+/Q+ homozygotes increase the trait above the overall mean, *m*, by *a* units while the Q2/Q2 homozygotes decrease the mean by *a*, then the means of the four gametic types are as shown in Table, where *R* is the recombination frequency between the QTL and the marker. From this it follows that the mean trait score of all those DH lines which are M1/M1 (M1M1) is *m* 1 *a*(1 2 2*R*), while the mean of all those DH lines which are M2/M2 (M2M2) is *m* 2 *a*(1 2 2*R*). In other words, half the difference between these two means (M1M1 2 M2M2) is *a*(1 2 2*R*).

Clearly, if the marker is so close to the QTL that it never recombines, *R* will be 0 and the marker difference will represent the QTL effect, *a*. Conversely if the two loci are unlinked, *i.e.* $R = 0.5$, then the difference will be zero. This effect is, of course, the basis of bulked segregant analysis. In general, the magnitude of the marker effect will decline linearly with the distance of the marker from the QTL in terms of recombination frequency, and so with several linked markers, their individual trait effects will be as shown in Figure. Similar Arguments Apply to Other Populations Such as F_2s and RILs.

If the trait means of the markers and their map are known very accurately, and only one QTL existed on the chromosome, then as Figure shows, it would be relatively easy to locate the QTL. Figure shows the distribution of flowering time among a number of doubled haploid lines of *Brassica oleracea* containing alternative alleles at a marker located very close to a QTL. In practice neither the map positions of the markers nor their means are known with very great accuracy because the number of DH lines and replicated plots are generally few and the heritability of quantitative traits is generally low. Moreover, linked QTLs create difficulties because their individual effects can combine either to create the impression of an intermediate, ghost QTL, or to conceal their effects altogether.

Various analytical procedures have been developed to maximize the efficiency and accuracy of locating individual QTLs and to separate linked QTLs. They all rely on the relationships described in the previous paragraph and use either maximum likelihood or weighted least squares regression procedures to estimate the QTL locations and effects that best fit the observed trait scores for each genotype.

The main difficulty is that there are two unknown parameters with respect to each marker score; the QTL effect, *a*, and map position. This means that it is necessary, in all methods, to use an iterative approach which involves trying all possible QTL locations along each chromosome and identifying the position or positions of the QTL which best fit the observed data as indicated by the size of the likelihood or the residual regression variance. As a result, a very large number of statistical tests are performed with the concomitant risk of false positive results.

All methods provide estimates of QTL locations and effects with similar precision in terms of confidence intervals, and this precision drops rapidly as the heritability of individual QTLs declines, *i.e.* the estimates are accurate only when there are just two or three unlinked QTLs with large effects. For example, Hyne *et al.* showed that the 95 per cent confidence interval for the location of a QTL contributing 10 per cent to the phenotypic variance of an F2 could be as large as 35 cM.

However, conservationists and breeders who are concerned with introducing QTL into commercial cultivars from other cultivars or more distantly related species, would normally only be interested in those cases where a few QTLs of major effect are to be manipulated. There is no evidence that dense maps are required for initial

QTL location. It is better to have large populations and a few (5–10) markers per chromosome. There is evidence that many of the larger QTLs located so far map closely to previously known major genes. This is often used as evidence that there are, in fact, very few QTLs, but this may well be misleading for several reasons. The QTL map locations are sufficiently imprecise that they could well appear spuriously close to candidate loci.

Only QTLs of large effect will stand much chance of being located and, by definition, they are likely to be major genes. It is likely that many cases where QTLs have very large effects could be due to the chance association of alleles of like effect at several QTLs along a chromosome, which cannot be separated as individual effects. Much effort has been devoted to increasing the precision of QTL location, in particular in improving the power of the test to reduce the failure to detect genuine QTL.

A long-established problem with breeding for quantitative traits in plants is the genotype 3 environment interaction. Using QTL location techniques, it is now possible to explore these effects at the level of individual QTLs although it is often difficult to know whether genotype 3 environment or poor repeatability is responsible.

CLONING AND CLONING VECTORS

A cloning vector is a small piece of DNA into which a foreign DNA fragment can be inserted. The insertion of the fragment into the cloning vector is carried out by treating the vehicle and the foreign DNA with the same restriction enzyme, then ligating the fragments together. There are many types of cloning vectors. Genetically engineered plasmids and bacteriophages (such as phage λ) are perhaps most commonly used for this purpose. Other types of cloning vectors include bacterial artificial chromosomes (BACs) and yeast artificial chromosomes (YACs).

Most commercial cloning vectors have key features that have made their use in molecular biology so widespread. In the case of expression vectors, the main purpose of these vehicles is the controlled expression of a particular gene inside a convenient host organism (eg. E. coli). Control of expression can be very important; it is usually desirable to insert the target DNA into a site that is under the control of a particular promoter. Some commonly used promoters are T7 promoters, lac promoters (bla promoter) and cauliflower mosaic virus's 35s promoter (for plant vectors).

To allow for convenient and favourable insertions, most cloning vectors have had nearly all their restriction sites engineered out of them and a synthetic multiple cloning site (MCS) inserted that contains many restriction sites. MCSs allow for insertions of DNA into the vector to be targeted and possibly directed in a chosen orientation. A selectable marker, such as an antibiotic resistance [eg. beta-lactamase (see figure)] is often carried by the vector to allow the selection of positively transformed cells (see Screening below). All plasmids must carry a functional origin of replication (ORI; not shown in figure). Some other possible features present in cloning vectors are: vir genes for plant transformation, intergrase sites for chromosomal insertion, lacZα fragment for α complementation and blue-white selection, and/or reporter genes in frame with and flanking the MCS to facilitate the production of recombinant proteins [eg. fused to the Green fluorescent protein (GFP) or to the glutathione S-transferase.

Properties of all Cloning Vectors

- Will accept foreign DNA and can still complete their life cycle
- Permit those bacterial or yeast cells that contain the foreign DNA to be distinguished from those that do not

Types of Vectors

- Plasmids.
- Bacteriophage.
- Cosmids.
- Yeast artificial chromosomes (YACs).
- Bacterial artificial chromosomes (BACs).

Insert Size of Various DNA Cloning Vectors

	Vector	Insert size (kb)
	Plasmid	<10 kb
	Bacteriophage	9-20 kb
	Cosmids	33-47 kb
	BACs	75-125 kb
YACS	100-1000 kb	

PLASMIDS

Characteristics of Plasmids

- Extrachromosomal circular DNA molecules which are not part of the bacterial genome.
- Size range: 1-200 kb.
- Carry functions advantageous to the host such as.
  - Produce enzymes which degrade antibiotics or heavy metals
  - Produce restriction and modifying enzymes
- Replication is coupled to host replication in a:
  - Stringent manner: One (or two) plasmids made during each round of bacterial replication
  - Relaxed manner: 10-200 copies of the plasmid made during each round of bacterial replication; (this can be increased to 1000-2000 plasmids by stopping host protein synthesis and replication with the antibiotic chloramphenicol)

Foreign DNA is inserted into a plasmid (or any cloning vector) by ligating the DNA into a complementary site in the plasmid. These sites are generated by digesting the DNA and vector with the same restriction enzyme. (The site for the restriction enzyme that is chosen should only be represented once in the plasmid.

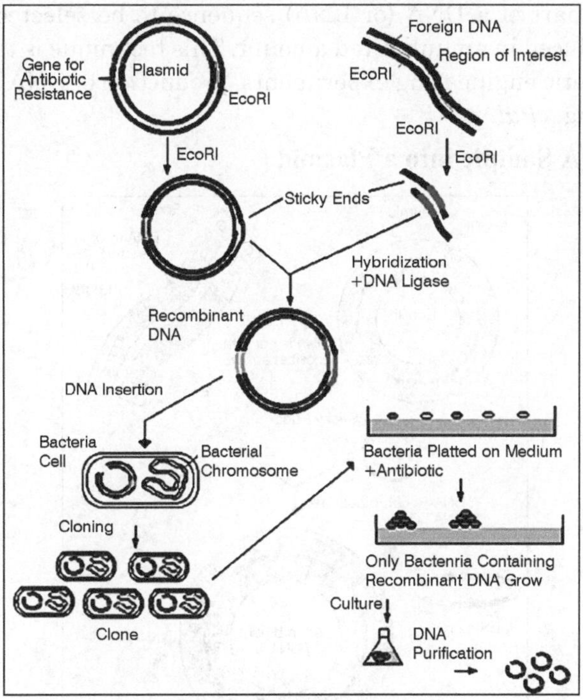

Fig. Cloning into a Plasmid

Thus, when the plasmid is digested, a single, linear molecule would be generated.) The foreign DNA is then inserted into the plasmid by the action of the enzyme DNA ligase. The next step is to insert the ligated DNA into a bacterial cell for propagation. This is done by a technique called transformation. Bacterial cells are treated with either Ca2Cl or Rb2Cl. This treatment produces pores in the bacterial cell wall and membrane through which the plasmid enters. Although there is no size limitation to the ligation reaction, transformation efficiency is dictated by the size of the plasmid. Process by which a plasmid is used to import recombinant DNA into a host cell for cloning. Many diseases are caused by gene alterations. Our understanding of genetic diseases was greatly increased by information gained from DNA cloning. In DNA cloning, a DNA fragment that contains a gene of interest is inserted into a cloning vector or plasmid.

The plasmid carrying genes for antibiotic resistance, and a DNA strand, which contains the gene of interest, are both cut with the same restriction endonuclease. The plasmid is opened up and the gene is freed from its parent DNA strand. They have complementary "sticky ends." The opened plasmid and the freed gene are mixed with DNA ligase, which reforms the two pieces as recombinant DNA Plasmids + copies of the DNA fragment produce quantities of recombinant DNA. This recombinant DNA stew is allowed to transform a bacterial culture, which is then exposed to antibiotics. All the cells except those which have been encoded by the plasmid DNA recombinant are killed, leaving a cell culture containing the desired recombinant DNA. DNA cloning allows a copy of any specific part of a DNA (or RNA) sequence to be selected among many others and produced in an unlimited amount. This technique is the first stage of most of the genetic engineering experiments: production of DNA libraries, PCR, DNA sequencing, *et al.*

Inserting a DNA Sample into a Plasmid

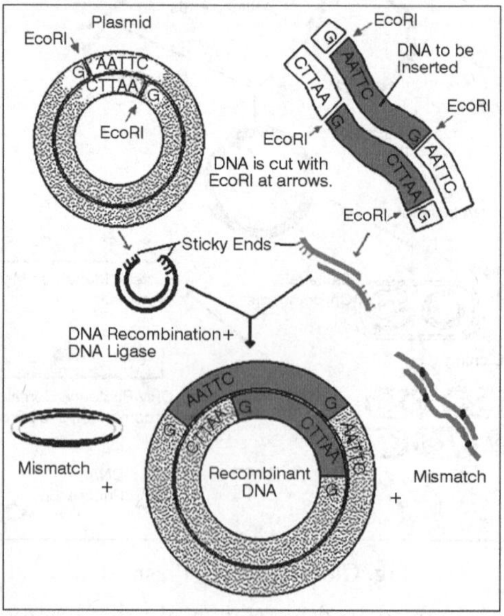

Fig. Inserting a DNA Sample in to a Plasmid

Plasmids are similar to viruses, but lack a protein coat and cannot move from cell to cell in the same fashion as a virus. Plasmid vectors are small circular molecules of double stranded DNA derived from natural plasmids that occur in bacterial cells. A piece of DNA can be inserted into a plasmid if both the circular plasmid and the source of DNA have recognition sites for the same restriction endonuclease.

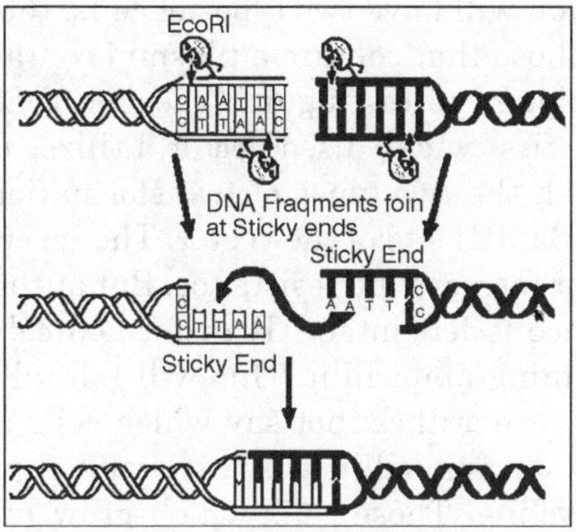

Fig. Restriction Enzyme Action of EcoRI

Restriction enzymes, also called restriction nucleases (EcoRI in this example), surrounds the DNA molecule at the point it seeks(sequence GAATTC). It cuts one strand of the DNA double helix at one point and the second strand at a different, complementary point (between the G and the A base). The separated pieces have single stranded "sticky-ends," which allow the complementary pieces to combine. The newly joined pieces are stabilized by DNA ligase. EcoRI, one of many restriction enzymes, is obtained from the bacteria Escherichia coli. The plasmid and the foreign DNA are cut by this restriction endonuclease (EcoRI in this example) producing intermediates with sticky and complementary ends. Those two intermediates recombine by base-pairing and are linked by the action of DNA ligase. A new plasmid containing the foreign DNA as an insert is obtained. A few mismatches occur, producing an undesirable recombinant. The new plasmid can be introduced into bacterial cells that can produce many copies of the inserted DNA. This technique is called DNA cloning.

Effect of Plasmid Size on Transformation Efficiency

Molecule size (kb)	% Maximum probability
2.0	57
3.2	100
4.3	86
12.5	43
20.0	36
39.0	14
54.0	6

The goal of the ligation reaction is to insert the foreign DNA into the vector. An unwanted ligation product also occurs - religated vector. One way to minimize this event is to treat the vector with phosphatase. This removes the terminal phosphate group from the restriction site of the vector and theoretically prevents religation of the two ends. In reality though, this treatment is never 100 per cent, so a low level of religation does occur. Thus after transformation you will have two types of cells: those which contain the original plasmid and those that contain a plasmid containing foreign DNA.

Plasmids are designed to distinguish the two types of transformation products. pBR322, the first widely used vector, utilizes differential antibiotic screening to distinguish the two types of transformation products. Let's say that we clone into the BamHI site of the vector. The insert DNA will then split the gene responsible for tetracycline resistance. But at the same time, the gene for ampicillin resistance is left intact. Transformed cells are first grown on bacterial plates containing ampicillin. This will kill all the cells that do not contain a plasmid. But we still cannot say which cells contain foreign DNA. Those cells that grew on ampicillin are then replica plated on plates with ampicillin and tetracycline. Those cells which grow in the presence of the ampicillin, but die under tetracycline selection contain plasmids which have foreign DNA inserts.

pBR322 was a breakthrough for molecular biology, but the double screening procedure was time consuming and could be subject to error. In 1981, a new series of plasmids were developed that permitted the identification of the foreign DNA containing cells in a single screening step. These are called the pUC plasmids. As with pBR322, ampicillin resistance is used as one selectable marker. The second marker is based on insertional inactivation of the E. coli lacZ gene. The wild type gene can hydrolyze a specific dye [X-Gal (5-bromo-4-indoyl-B-D- galactopyranoside)] to a blue colour, and the bacterial colony is stained blue. A multiple cloning site has been inserted into this gene. This site will accept fragments ending in a number of different restriction enzymes. Upon insertion of DNA into this site, the activity of the gene is eliminated and the colony appears white in colour. Thus transformed colonies containing plasmids with inserts can be distinguished from those with plasmids without inserts based on the colour of the colony and the ability of the colony to grow on an ampicillin containing media.

cDNA Cloning (Cloning Eukaryotic mRNA)

cDNA cloning is a method of obtaining a DNA copy of the mRNAs that are expressed at a specific stage in the development of the plant. In this manner, you can enrich the library that you will screen for those sequences in which you are interested. The reagent required for this type of cloning approach is mRNA. These mRNAs have a poly A+ tail. This tail permits the isolation of poly A+ mRNA by using either oligo-dT or oligo-U columns. Total RNA is run through one of these columns under conditions which favour the binding of the tail to the matrix on the column. After the column is extensively washed, the conditions are changed and the bound mRNA is isolated. This is the starting reagent for cDNA cloning.

Steps in cDNA Cloning

- Bind oligo-dT to the poly +A tail of the mRNA.
- Add reverse transcriptase and make a DNA copy of mRNA (cDNA). Erase the mRNA with alkali and high temperature. (First strand synthesis)
- Add a C-tail to the 3' end of the cDNA with terminal transferase.
- Add oligo-dG to the tailed-cDNA and make the second strand with reverse transcriptase or the Klenow fragment of DNA Polymerase I.
- Add dC's to the 3' end of the double-stranded cDNA with terminal transferase
- Add C-tailed, ds-cDNA to G-tailed, PstI cut pBR322 and anneal. (DNA ligase is not required for this step.)
- Transform E. coli cells.
- Select Tetr/Amps cells. Most, but not all inserts can be removed by PstI digestions.

BACTERIOPHAGE LAMBDA VECTORS

We have talked about plasmids as vectors for cloning small pieces of DNA. The limitation of this vector is the size of DNA that can be introduced into the cell by transformation. This presents problems when you are trying to create a genomic library of a large genome such as with plants. A genomic library contains all of the DNA found in the cell of the plant (or any organism). If you digest plant DNA to completion with a restriction enzyme, ligate those fragments into a plasmid vector and transform bacterial cells, only a portion of those fragments will be represented in the final transformation products. If a gene of interest is located on a large fragment then you will not be able to isolate that gene from a plasmid library.

But what can be done to increase the probability of obtaining a clone which contains the entire gene. First you need to use a vector that can accept large fragments of DNA. Examples of these are bacteriophage and cosmid vectors and more recently yeast artificial chromosomes. Bacteriophage lambda vectors were developed because several observations were made that suggested that they could complete their life cycles even if foreign DNA was inserted into a portion of its genome. This suggested that certain regions of the virus were not essential. Let's first discuss the life cycle of lambda.

- *Adsorption*: The phage particle binds at a maltose receptor site of the bacterial cell; growing the cell in the presence of the sugar increase the number of receptor sites
- *Penetration*: DNA is injected into the cell; at this point it can enter one of two pathways;

- *Lysogenic pathway*: The phage DNA becomes integrated into the genome and is replicated along with the bacterial DNA; it remains integrated until it enters the lytic pathway
- *Lytic pathway*: Large scale production of bacteriophage particles that eventually leads to the lysis of the cell; base pairing at the cos site leads toa circular molecule.

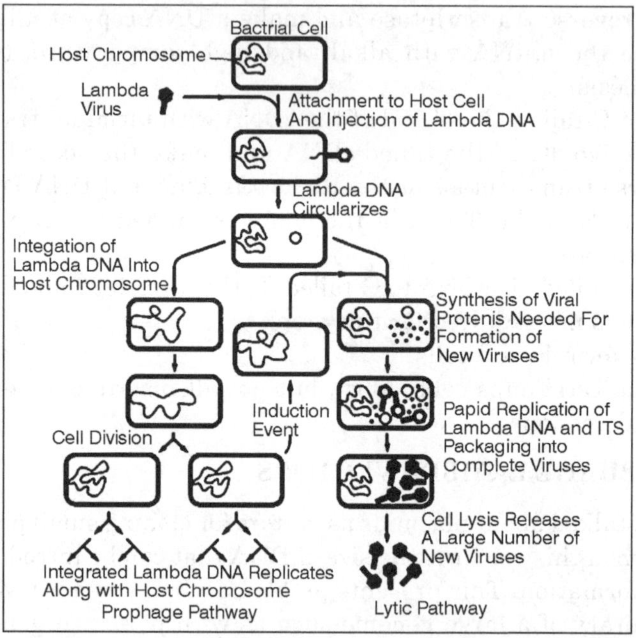

The linear double-stranded DNA lambda genome contains about 50,000 nucleotide pairs and encodes 50-60 different proteins. When the lambda DNA enters the cell the ends join to form a circular DNA molecule. The bacteriophage can multiply in E. coli by a lytic pathway, which destroys the cell, or it can enter a latent prophage state. Damage to a cell carrying a lambda prophage induces the prophage to exit from the host chromosome and shift to lytic growth (green arrows). The entrance and exit of the lambda DNA from the bacterial chromosome are site-specific recombination events

- *Early transcription*: Transcription proceeds from the pL and pR promoters, through the N and cro genes and stops at terminators tL and tR1; a low level of transcription through the O and P genes occurs and terminates at tR2; the N product is an antitermination factor that is important for the next stage of transcription.
- *Delayed early transcription*: The N product binds to RNA polymerase and transcription proceeds past the tL, tR1 and tR2 terminators; genes to the left of N, involved in recombination, to the right of cro, involved in replication, are expressed at this point; another protein expressed from the Q gene is used for antitermination of later transcription.

- *Replication*: Early replication is through a theta form initiated from a single origin of replication site; later replication is via rolling circle replication; this produces long concatamers of the phage DNA that are cleaved at the cosL and cosR sites.
- *Late transcription*: The protein product of the cro gene builds up to a critical level and then binds to the oL and oR to stop early transcription; another protein, a product of the Q gene, has built up and activates transcription at the p'R promoter by antitermination; transcription terminates with in the b region; this transcription results in the production of the proteins required for the head and tail of the mature phage particle and those required for bacterial cell lysis.
- *Assembly*: A prophage head is produced; a unit length DNA is placed into the head by the action of the Nu1 and A proteins; the DNA is locked into place by the D protein and ter function of the A protein clips the DNA at the cosL and cosR sites; the concatamer is released, the tail is added and the mature phage particle is completed.

Packaging of the DNA into the head does not require a complete length of wild type lambda. It has been determined that a lambda molecule that is between 78 per cent and 105 per cent of wild type length can be packaged. This is from 37 to 53 kb in length.

Two important developments suggested that lambda may be suitable as a cloning vector. First it was determined that the gene products between the J and N genes could be removed and the life cycle could be completed. Second, restriction enzyme sites could be eliminated which permitted the development of a vector with a single site for insertion of foreign DNA.

Two types of vectors have been developed:

- *Insertional vector*: DNA is inserted into a specific site.
- *Replacement vector:* Foreign DNA replaces a piece of DNA (stuffer fragment) of the vector.

Let's talk about a specific vector EMBL 3 and EMBL 4. One important concern when cloning with lambda vectors is that you want to maximize the number of resulting phage particles that contain foreign DNA. Or said another way you want to minimize the number of wild type particles. One approach is through spi selection. This refers to sensitivity to P2 interference.

Table. Bacteriophage Phenotypes for Growth on P2 Lysogens

Phenotype	Growth on P2 Lysogens? (bacterial strain)
spi+ (red+gam+)	Poor
spi– (red–gam–)	Good

EMBL 3/4 vectors have placed the red and gam genes in the stuffer fragment. Thus only those particles from which the stuffer has been replaced can grow well in a P2 lysogen bacterial cell.

Cloning in Lambda Vectors

Phages are viruses that can infect bacteria. The major advantage of the phage vector is its high transformation efficiency, about 1000 times more efficient than the plasmid vector.

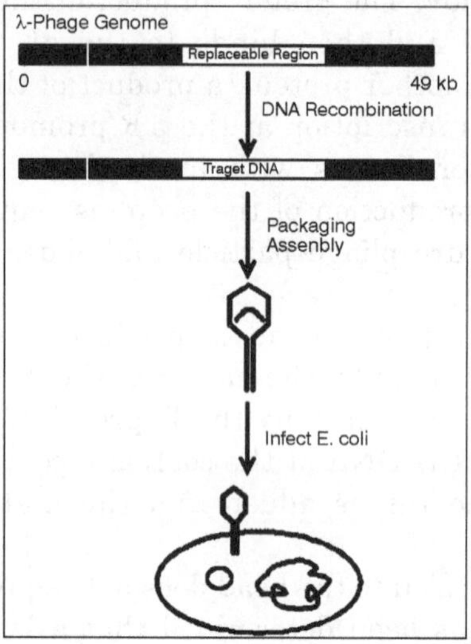

Fig. Schematic Drawing of the DNA Cloning Using ? Phages as Vectors.

The DNA to be cloned is first inserted into the DNA, replacing a non-essential region. Then, by an in vitro assembly system (described below), the virion carrying the recombinant DNA can be formed. The genome is 49 kb in length which can carry up to 25 kb foreign DNA.

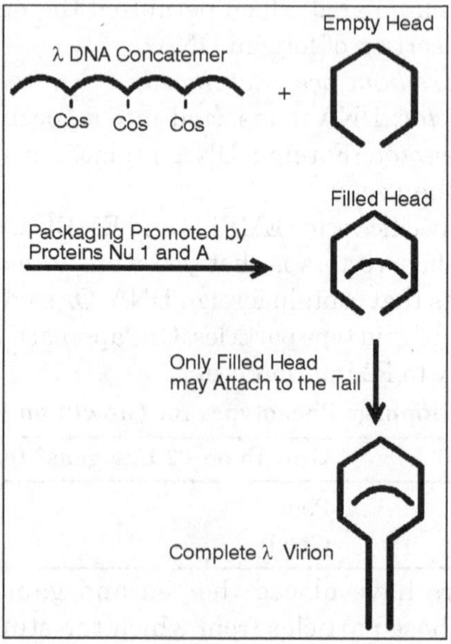

Fig. The Assembly Process of the Virion.

The extreme ends of the DNA are known as COS sites, each is single stranded, 12 nucleotides long. Because their sequences are complementary to each other, one end of DNA may base-pair with the other end of a different DNA, forming concatemers. The two ends of a DNA may also bind together, forming a circular DNA. In the host cell, the DNA circularizes because ligase may seal the join of the COS sites. In the assembly process of virions, two proteins Nu1 and A can recognize the COS site, directing the insertion of the DNA between them into an empty head.

The filled head is then attached to the tail, forming a complete virion. The whole process normally takes place in the host cell. However, to prepare the virion carrying recombinant DNA, the following in vitro assembly system is commonly used. Proteins Nu1 and A are encoded by the genes in the genome. If the two genes are mutated, DNA cannot be packaged into the pre-assembled head. Because tails attach only to filled heads, the cell will accumulate separate empty heads and tails, which can then be extracted. When the extract is mixed with recombinant DNA and proteins Nu1 and A, the complete virion carrying recombinant DNA will be assembled.

COSMIDS

Cosmids are plasmid vectors that contain cos sites. The cos site is the only requirement for DNA to be packaged into a phage particle. Cosmids were developed in light of this observation. How do you clone into cosmid vectors?

- Clone the DNA into the vector as you would with any plasmid.
- Introduce the DNA into the bacterial cell via a phage particle.
- Propagate as plasmid.

Since phage particles can accept between 38 and 53 kb of DNA and since most cosmids are about 5 kb, between 33 and 48 kb of DNA can cloned in these vectors.

It has the following advantages:

- High transformation efficiency.
- The cosmid vector can carry up to 45 kb whereas plasmid and phage vectors are limited to 25 kb.

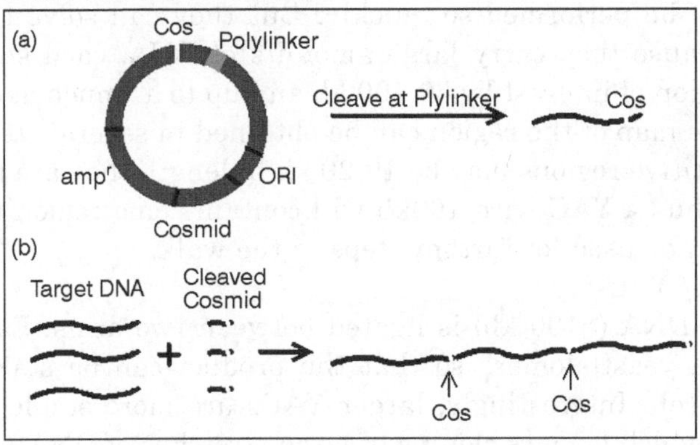

Fig. Cloning by using Cosmid Vectors.

(a) In Addition to Ampr, ORI, and Polylinker as in the Plasmid Vector, the Cosmid Vector also Contains a COS Site. (b) After Cosmid Vectors Are Cleaved with Restriction Enzyme, they are Ligated with DNA Fragments. The Subsequent Assembly and Transformation Steps are the same as Cloning with ? Phages.

Problems Associated with Lambda and Cosmid Cloning

- Since repeats occur in eukaryotic DNA rearrangements can occur via recombination of the repeats present on the DNA inserted into lambda or cosmid.
- Cosmids are difficult to maintain in a bacterial cell because they are somewhat unstable.

YEAST ARTIFICIAL CHROMOSOMES (YACS)

One goal of molecular genetics is to obtain physical data about the genomic organization of long stretches of DNA. Traditionally, this data has been obtained by a technique called chromosome walking. Walking is performed by subcloning the end of a lamda or cosmid clone and screening a library for other clones that contain similar sequence information. If this new clone overlaps a portion of the original clone, then the length of the DNA of interest is extended by the length of DNA in the second clone that is not found in the original clone.

By performing these steps successive times, a long distance map can be obtained. This technique though has difficulties. First, each step is technically slow. Second, if you use lambda or cosmid clones, you might only extend the region of interest by 5-10 kb in each step of the walk. Finally, if any of the clones that are obtained contain repeated sequences, the subclone could lead you to another region of the genome that is not contiguous with the region of interest.

Yeast artificial chromosomes can alleviate some of these problems because of the large (100-1000kb) amount of DNA that can be cloned. First of all, YACs cannot speed up each step of the walk because the subcloning and screening steps can only be performed so quickly. But they can solve the other two problems. Because they carry large amounts of DNA, each step can easily extend the region of interest by 50-100 kb and up to as much as 500 kb. Thus a long distance map of the region can be obtained in several steps. Secondly, although repetitive regions may be 10-20 kb in length they are rarely, longer than 50 kb. Thus a YAC with 100kb will contain some region that is single copy which can be used for further steps in the walk.

Features of YACs:

- Large DNA (>100 kb) is ligated between two arms. Each arm ends with a yeasttelomere so that the product can be stabilized in the yeast cell. Interestingly, larger YACs are more stable than shorter ones, which favours cloning of large stretches of DNA.

- One arm contains an autonomous replication sequence (ARS), a centromere (CEN) and aselectable marker (trp1). The other arm contains a second selectable marker (ura3).
- Insertion of DNA into the cloning site inactivates a mutant expressed in the vector DNA and red yeast colonies appear.
- Transformants are identified as those red colonies which grow in a yeast cell that is mutant for trp1 and ura3. This ensures that the cell has received an artificial chromosome with both telomeres (because of complementation of the two mutants) and the artificial chromosome contains insert DNA (because the cell is red).

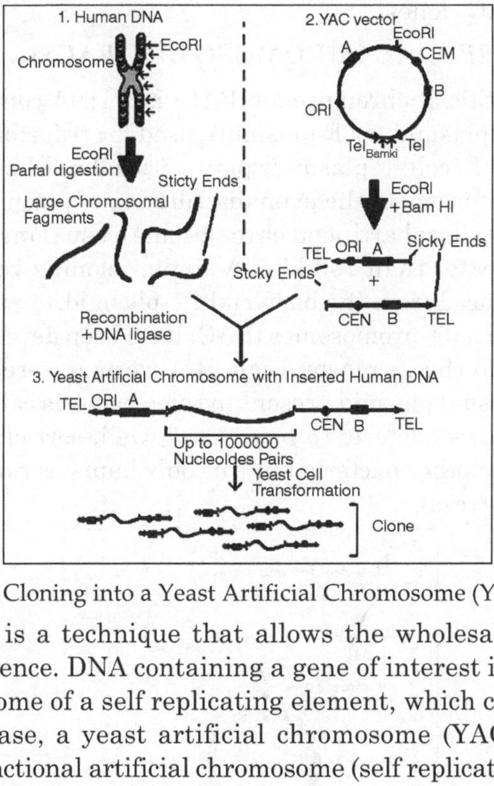

Fig. Cloning into a Yeast Artificial Chromosome (YAC)

DNA cloning is a technique that allows the wholesale production of a specific DNA sequence. DNA containing a gene of interest is inserted into the purified DNA genome of a self replicating element, which can be a plasmid, a virus or in this case, a yeast artificial chromosome (YAC). A YAC can be considered as a functional artificial chromosome (self replicating element), since it includes three specific DNA sequences that enable it to propagate from one cell to its offspring:

- *TEL*: The telomere which is located at each chromosome end, protects the linear DNA from degradation by nucleases.
- *CEN*: The centromere which is the attachment site for mitotic spindle fibres, "pulls" one copy of each duplicated chromo-some into each new daughter cell.
- *ORI*: Replication origin sequences which are specific DNA sequences that allow the DNA replication machinery to assemble on the DNA and move at the replication forks.

It also contains few other specific sequences like:

- *A and B*: selectable markers that allow the easy isolation of yeast cells that have taken up the artificial chromosome.
- Recognition site for the two restriction enzymes EcoRI and BamHI.

While DNA cloning into a plasmid allows the insertion of DNA fragment of about 10,000 nucleotide base pairs, DNA cloning into a YAC allows the insertion of DNA fragments up to 1,000,000 nucleotide base pairs. Why is it so important to be able to clone such large sequences? To map the entire human genome (3x1,000,000,000 nucleotide base pairs) it would require more than 100,000 plasmid clones. In principle, the human genome could be represented in about 10,000 YAC clones.

BACTERIAL ARTIFICIAL CHROMOSOMES (BACS)

A bacterial artificial chromosome (BAC) is a DNA construct, based on a functional fertility plasmid (or F-plasmid), used for transforming and cloning in bacteria, usually E. coli. F-plasmids play a crucial role because they contain partition genes that promote the even distribution of plasmids after bacterial cell division. The bacterial artificial chromosome's usual insert size is 150-350 kbp, but can be greater than 700 kbp. A similar cloning vector called a PAC has also been produced from the bacterial P1-plasmid.

Bacterial Artificial Chromosomes (BAC) have been developed to hold much larger pieces of DNA than a plasmid can. BAC vectors were originally created from part of an unusual plasmid present in some bacteria called the F' plasmid. The F' plasmid allows bacteria to have "sex" (well, sort of: F' helps bacteria give its genome to another bacteria but this only happens rarely when bacteria are under a lot of stress).

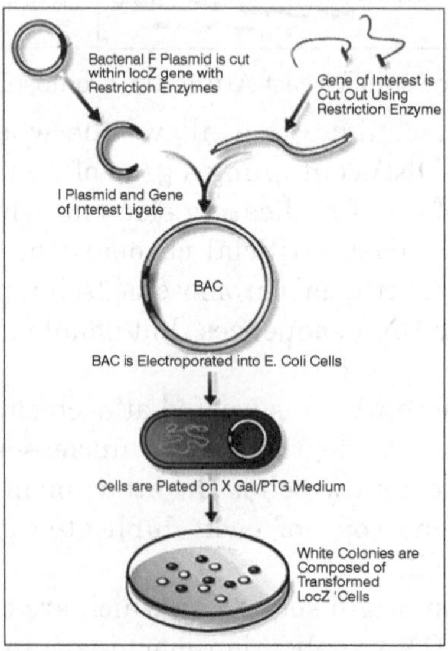

Fig. Transforming a Bacterium Using a BAC Vector

F' had been studied extensively and it was found that it could hold up to a million basepairs of DNA from another bacteria. Also, F' has origins of replication and bacteria have a way to control how F' is copied. In 1992, Hiroaki Shizuya took the parts of F' that were important, cleaned it up, and turned it into a vector. BAC vectors are able to hold up to 350 kb of DNA and have all of the tools that a vector needs to work properly, like replication origins, antibiotic resistance genes, and convenient places where clone DNA can insert itself. With these vectors it is possible to study larger genes, several genes at once, or entire viral genomes. By using a vector that can hold larger pieces of DNA, the number of clones required to cover the human genome six times theoretically could drop from 1.8 billion to about 50 million.

Researchers have modified BAC vectors to become more convenient to use and more useful in specialized situations. In addition to the antibiotic resistance gene that was added to identify transfected bacteria, a gene was added that enabled the bacteria to turn the colourless substance X-gal/IPTG blue. This substance is found in the chicken-soup-like media on which the bacteria colonizes. This colour-changing gene, called lacZ, is split apart when the clone DNA is incorporated into the vector, so it is possible to tell not only if a bacteria had been transfected (meaning incorporated into the cell), but also if the bacteria was transfected with the vector containing insert DNA or just the vector alone (remembering that if the vector has properly incorporated the clone DNA, it will have lost its ability to change X-gal/IPTG).

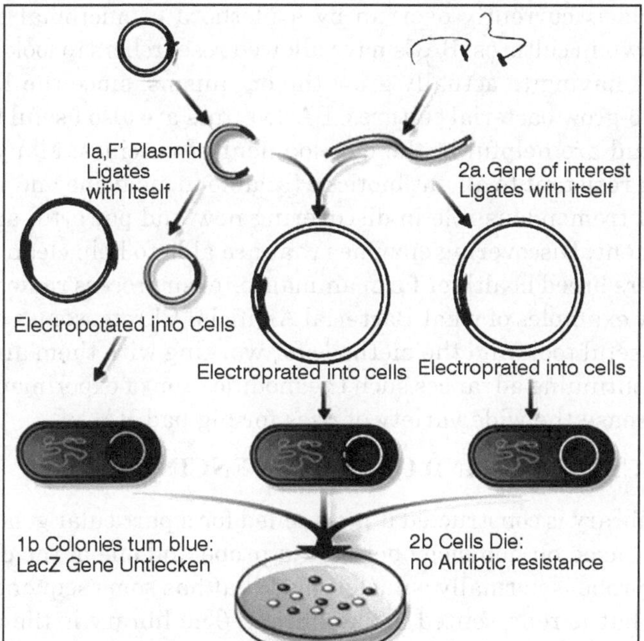

Fig. Selecting for Transformed Bacteria

Another similar modification uses the gene called sacB, which encodes a protein called levansucrase. This protein turns sucrose - table sugar - into levan, a toxic substance to bacteria. In a similar fashion, bacteria grown on media with sucrose will die if sacB is not broken up by inserted DNA. If the vector carries a DNA insert, then sacB is broken and it won't produce levansucrase, so the bacteria can survive in the sucrose media. In theory, only bacteria transfected with a vector containing insert DNA would be able to grow and form colonies.

Some modifications to BAC vectors make them more specialized. For instance, there are researchers studying the herpes virus who have made a BAC vector that can be cultured in bacterial cells and then when put into a mammalian cell, instantly releases its insert DNA - in this case, the whole herpes virus genome. With such a vector, it is easier to grow sufficient amounts of the herpes virus for research, since it can live in bacterial cultures, instead of requiring their endemic mammalian cell cultures, which are extremely difficult to maintain. This in turn makes it easier to make modifications to the virus and study what each of its genes do. The creation of BAC vectors has allowed researchers to do many things that they could not do before and do them more quickly and more easily. Some of the scientific practices that have simplified with BAC vectors includes phylogenetic studies (studies which examine species' relationship to one another) and what the absolute minimum size of a genome could be.

The world is currently overrun by a plethora of microbial species that cannot be grown in cultures. BACs have allowed researchers to look at microbial DNA without having to actually grow the organisms, since the DNA is kept within easy-to-grow bacterial cultures. BAC vectors are also useful for studying pathogens, and are helpful in the development of vaccines. Many pathogens are becoming resistant to all antibiotics available to medicine and BAC vectors are playing a tremendous role in discovering new and powerful antibiotics in the environment. Discovering enzymes that are able to help clean up oil spills or help farmers breed healthier farm animals or even process radioactive waste are just a few examples of what Bacterial Artificial Chromosomes can do. BAC vectors are useful tools and the methods for working with them are fairly well developed. Continuing advances such the modification of experimental animals will only increase the wide variety of uses for big bad BACs.

LIBRARY SCREENING AND GENE SEQUENCING

Once a library is constructed it is screened for a particular gene of interest. Screening is based on homology between a probe and one of the clones in the library. The probe is normally a nucleic acid that has some sequence homology to the gene that is represented in the library. The library is the collection of clones from the source DNA which is inserted into a cloning vector. For example, a bean genomic lambda library contains pieces of the entire complement of bean DNA, from 9-20 kb in size, cloned into a lambda vector. Any probe used to screen the library should have some homology to a clone in the bean library, and this homology would allow you to select the appropriate clone from the many clones that do not contain your DNA of interest.

Probes come in several forms. The more homologous (similar in DNA sequence) the probe is to the sequence that is being sought, the easier it is to select a clone from the library. For example, a bean lambda library is considered a genomic library because it contains all the DNA sequences found in the bean genome. A bean leaf cDNA library, though, would contain only those sequences that are expressed in the leaf after they have undergone processing.

Thus the cDNA clones would not contain any of the intron sequences or the controlling elements of the gene. To obtain a genomic clone from the bean lambda library, the best probe would be a cDNA clone obtained from screening a bean cDNA library. This type of probe, a probe that contains the exact sequence of the sequence that is being sought, is called a homologous probe. But how was the original cDNA clone obtained so that it could be used as a probe.

The cDNA library from which the clone was obtained might have been screened by a probe from another species which represents coding information for the same gene but from another species.

For example, a bean leaf cDNA library may be screened with a tomato RUBISCO small subunit clone. A probe that contains DNA which encodes for the gene of interest but from another species is called a heterologous probe. Many plant genes have been cloned by screening libraries with both homologous and heterologous probes.

Polymerase chain reaction techniques are now commonly used to clone genes. To use this technique primers must be designed that are complementary to your target sequence. One oligonucleotide will be complementary to the anticoding strand (the DNA strand of the gene complementary to the mRNA and used as a template for transcription). The second oligonucleotide will be complementary to the coding strand (the DNA strand of the gene complementary to the anticoding strand).

If the gene has been cloned in another species, you can use that sequence information to design primers to amplifiy a fragment of the gene from the DNA of your species. Often, the gene you have an interest in has never been cloned, but you may have isolated the protein. If this is the case, the first step of this method is to obtain partial protein sequence information of the protein. Micropeptide sequencers are available that can rapidly generate sequence information.

From this sequence, you can use reverse translation of the amino acid sequence to obtain the nucleic acid sequences of an amino-terminal fragment and a carboxy-terminal fragment. In this case the fragment complementary to anticoding strand will be a direct conversion of the genetic code for the amino acids in the amino-terminal fragment of the protein. The strand complementary to the coding strand will be complementary to the derived nucleic acid sequence of the carboxy-terminal fragment. Let's say that the following is the N-terminal sequence of the peptide for which you want to derive a synthetic oligonucleotide:

NH_3+–Met–Cys–Val–Lys–Thr–Pro–CO_2–

The following would be an appropriate probe for the gene.

5'-A T G T G T/G G T N A A A/G A G N C C – 3'

(N = A, T, G, and C)

When you are constructing these probes several concepts should be kept in mind. First, the genetic code is written in the mRNA sequence. This synthetic oligonucleotide will thus, be complementary to the anit-coding strand that is used as the template for the mRNA. Secondly, because you only know the amino acid and not the DNA sequence you must deal with redundancies of the genetic code.

For example leucine can be represented by CTA, CTT, CTG, CTC. Thus your synthetic oligonucleotide needs to contain all the possibilities. Therefore the above sequence is actually a mixture of 64 (4×4×2×2) different sequences. Because of this redundancy it is always best to make you oligonucleotide one short of full length.

This will help some of the redundancy problems. In this example, the probe is 17 nucleotides long and the redundancies for proline are eliminated. Another method of reducing the redundancy problem is to incorporate the nucleotide deoxyinosine at any position where a high level or redundancy occurs. Deoxyinosine is a modified nucleotide that does not pair with any of the four bases found in DNA. Therefore, it does not affect the homology of the probe.

The following may be the amino acid sequence of a carboxy-terminal fragment:

NH_3+–Phe–Tyr–His–Thr–Val–Ala–CO_2–

The appropriate oligonucleotide sequence from the above sequence for PCR amplification would be:

5'– G C N A C N G T A/G T G A/G T A A/G A A -3'

Amplification of DNA between these two oligonucleotides will provide a homologous probe because it will represent the exact sequence of the gene for which you are searching. Once this probe is radiolabelled by nick-translation (as all probes must be radilabelled before library screening), it can be used to screen a cDNA or genomic library for a clone complementary to the amplified fragment.

DNA Sequencing With the Chain-Termination, Dideoxy Technique of Sanger

Once you have a candidate clone, you will want to sequence it to determine if it is similar to another gene already sequenced or whether it is unique. DNA sequencing is a DNA replication based reaction. The two requirements for DNA replication are a DNA template and a free 3'-OH group. These requirements must be met for any sequencing procedure. The following steps illustrate the Sanger procedure, the most widely used DNA sequencing procedure.

- Although this technique originally used DNA cloned in a vector that can generate single- stranded DNA, sequencing is now routinely performed in double-stranded plasmid or cosmid vectors. pUC plasmids and related plasmids are popular for double stranded sequencing. To begin the procedure, the DNA is made single stranded by a combination of heating and alkaline conditions.
- A primer which has a free 3'-OH group is annealed to the vector. A convenient primer is the multiple cloning site from the pUC plasmids. DNA synthesis will initiate here. 3. Four reactions are performed, each containing dATP, dGTP, TTP and 32P-dCTP and one dideoxy nucleotide in low concentration. Dideoxy nucleotides have a hydrogen rather than a - OH group attached to the #3 carbon and cannot be used for further extension of the chain. Thus in this reaction, various synthesis products will end with that nucleotide.
- Add DNA polymerase to the reaction. Nucleotides will be added one at a time but the chain will terminate when a dideoxy nucleotide is inserted. Because the dideoxy nucleotide is limiting a series of products, each ending with the same base, will be obtained.
- Separate the fragments on an acrylamide gel, develop via autoradiography and read the sequence.

5

Genetically Modified Crops

Global agriculture finds itself engrossed in a heated debate over genetically modified (GM) crops. This debate, which features science, economics, politics, and even religion, is taking place almost everywhere. It is going on in research labs, corporate boardrooms, legislative chambers, newspaper editorial offices, religious institutions, schools, supermarkets, coffee shops, and even in private homes. What is all the fuss about and why do people feel so strongly about this issue? This Pocket "K" attempts to shed light on the controversy by addressing several basic questions about GM crops.

APPLICATION OF RECOMBINANT DNA TECHNOLOGY

The application of recombinant DNA technology has revolutionized plant breeding in recent years. Genetic transformation in plant breeding greatly increases the gene pool and broadens the scope of genetic changes that modern breeders may draw upon. The first generation of genetically modified (GM) crop plants hold great promise in carrying new or improved agronomical traits that require less intensive farming methods or which have improved quality traits.

These GM crop plants have been produced to improve farming production systems, aiming, for instance, to reduce negative impacts on the environment, such as fertilizers, herbicides and pesticides. In general they are of little direct interest to the consumer. However, the genetic manipulation of the primary and secondary metabolism of plants offers distinct possibilities in the development of nutritionally improved products, raw materials with added pharmaceutical value, and crops with real health benefits for the consumer.

These are second-generation GM plants. Recently, several prototypes in molecular plant breeding with novel traits for (non-) food and industrial applications have been introduced into the market, and many more GM crops have the prospect of commercial exploitation in the near future. The safety and wholesomeness of these GM crop plants raises questions beyond those posed by conventional foods. In this context, traditional plant breeding of most food crops has conformed to the standard 'generally recognized as safe' (GRAS).

In some instances safety is assumed to be assured during performance and quality testing by breeders (*e.g.* glucosinolates and erucic acid in rape seed).

However, the process is largely focused on the comparison of agronomic and phenotypic characteristics, such as colour, maturity, yield or disease resistance. Products with an unusual taste or that have harmful effects following ingestion or skin contact have normally been rejected from the breeding programme. However, the safety of conventional plant varieties has usually been presumed from the assumption that prudent consumers will avoid those species once it is known that they cannot consume them without adverse health effects.

When the commercial production of GM plants started, it sparked a debate centred on issues associated with concerns on environmental, human and animal safety matters. Many people questioned whether the 'simple' GRAS approach, based on historical and empirical experiences, is sufficient to guarantee food safety of the GM crop plants. It is also asked whether the risks of chronic exposure of humans and animals to GM crops are adequately covered, and whether large-scale breeding of GM plants will have detrimental effects on ecosystems.

EVALUATION OF GENETICALLY MODIFIED FOOD CROPS

Genetic modification, otherwise referred to as recombinant DNA (rDNA) technology or gene-splicing, has proven to be a more precise, predictable and better understood method for the manipulation of genetic material than previously attained through conventional plant breeding. To date, agricultural applications of the technology have involved the insertion of genes for desirable agronomic traits (*e.g.* herbicide tolerance, insect resistance) into a variety of crop plants, and from a variety of biological sources. Examples include soybeans modified with gene sequences from a *Streptomyces* species encoding enzymes that confer herbicide tolerance, and corn plants modified to express the insecticidal protein of an indigenous soil microorganism, *Bacillus thuringiensis* (Bt). A growing body of evidence suggests that the technology may be used to make enhancements to not only the agronomic properties, but the food, nutritional, industrial and medicinal attributes of genetically modified (GM) crops.

Regulatory supervision of rDNA technology and its products has been in place for a longer period of time in the United States than in most other parts of the world. The methods and approaches established to evaluate the safety of products developed using rDNA technology continue to evolve in response to the increasing availability of new scientific information. As our understanding of the potential applications of the technology is broadened, the safety of products developed using rDNA technology and the potential effects of introduced gene sequences on human health or the environment will be more closely scrutinized. In fact, much of the knowledge acquired during the commercialization of the products of rDNA technology in agriculture is now finding application in evaluating the safety of products developed through more conventional means.

The objective of this chapter is to provide the reader with an overview of the significant events leading up to the present, science-based, regulatory framework that exists for the safety evaluation of GM food crops within the United States. An attempt has been made to discuss concerns over the sufficiency of existing regulations, as well as to highlight recent initiatives taken by federal regulatory agencies to address them. Through better communication of how the regulatory process functions within the United States, it is anticipated that current and future applications of rDNA technology in agriculture will be met with a greater level of understanding and acceptance.

HISTORICAL PERSPECTIVE

rDNA technology was first developed in the 1970s. The initial response of the scientific community, including members of the National Academy of Science (NAS), to the prospects of rDNA technology, was to postpone any further research involving the technology until the potential risks to human health and the environment could be evaluated.

Researchers attending the International Conference on Recombinant DNA Molecules in 1975, otherwise known as the Asilomar Conference, tried to establish a scientific consensus on how best to self-regulate emerging applications of the technology. The conditions and restrictions that were proposed at this conference have formed the basis by which federal guidelines and policies for rDNA technology research were drafted within the United States.

National Institutes of Health (NIH)

The National Institutes of Health (NIH) was the first federal regulatory agency to publish their interests in evaluating the safety of rDNA technology in 1976, in the form of guidelines for the conduct of research. Because of the uncertainties that existed at the time, all research into the potential applications of rDNA technology was limited to the confines of federally funded labouratories under NIH control. After continued research, and a more careful assessment and monitoring of the risks, a set of less restrictive guidelines was published in 1978.

However, the environmental release of organisms developed using rDNA technology outside the confines of controlled labouratory conditions was prohibited unless otherwise approved by the NIH director. In the early 1980s, the NIH established an rDNA Advisory Committee (RAC) to review all data and experience gained with applications of the technology under its control. Based on recommendations of the RAC, a more relaxed set of research guidelines was published by the NIH in 1983.

The NIH approved the first environmental release of an organism developed using rDNA technology (ice-minus strain of *Pseudomonas*) in 1983.

In response, they were criticized for failing to prepare a statement or assessment of the environmental impact of their regulatory decision as required under the National Environmental Policy Act (NEPA). Once the legal controversy had subsided, all responsibility that the NIH had for regulating the environmental introduction of GM organisms was relinquished. Nevertheless, NIH guidelines continue to be referenced in assessing the safety of rDNA research performed within industry, federal and other state labouratories. However, it was unclear which federal regulatory agencies would be responsible for ensuring the safety of the products developed using rDNA technology.

Office of Science and Technology Policy (OSTP)

In response to a need for clarification, the Office of Science and Technology Policy (OSTP) began work on the development of a policy to establish a federal regulatory framework for evaluating the safety of products developed using rDNA technology. Following an opportunity for public comment, OSTP published a final version of the 'Coordinated Framework for Regulation of Biotechnology' (Coordinated Framework) in 1986.

Table: Overview of Responsible Agencies under the Coordinated Framework

Responsible agency	Products regulated	Reviews for safety
FDA	Food, feed, food additives, veterinary drugs	Safe to eat
USDA	Plant pests, plants, veterinary biologic	Safe to grow
EPA	Microbial/plant pesticides, new uses of existing pesticides, novel microorganisms	Safety of a new use of a companion herbicide
	Safe for the environment	

The policy provided the basis by which federal regulatory agencies got involved in evaluating the safety of products at later stages of commercial development at that time.

According to the Coordinated Framework, the products of rDNA technology should be regulated on the basis of the unique characteristics and features that they exhibit, not their method of production. The products of rDNA technology were considered to pose risks to human health and the environment similar to those posed by conventional products already regulated within the United States. As a result, no new federal regulatory agencies or regulations were required. The Coordinated Framework did not, however, rule out the possibility of the development of new guidelines, procedures, criteria or even regulations to supplement or alter the scope of existing statutes for the products of rDNA technology.

The Coordinated Framework identified three federal regulatory agencies within the United States: the US Food and Drug Administration (US FDA), the US Department of Agriculture (USDA) and the US Environmental Protection Agency (US EPA), as having primary responsibilities for evaluating the products of rDNA technology under development at that time.

In 1992, the OSTP released another document entitled, 'Exercise of Federal Oversight within the Scope of Statutory Authority: Planned Introductions of Biotechnology Products Into the Environment', outlining the proper basis by which federal regulatory agencies were expected to exercise their regulatory authority. As with conventional products, dependent upon the intended use and function, more than one federal regulatory agency may share an interest in evaluating the safety of a product developed using rDNA technology. If more than one federal regulatory agency has an interest, lead agencies are identified as being responsible for coordinating activities to limit any potential duplication of efforts. Although federal regulatory agencies worked independent of one another, it was realised that close working relationships would need to be established in order to evaluate effectively the safety of products developed using rDNA technology.

Recently, the OSTP teamed up with the White House Council on Environmental Quality (CEQ) to perform a six-month inter-agency evaluation of the federal regulatory agency responsibilities in evaluating the environmental safety of products developed using rDNA technology. A case-study approach for a variety of different classes of products developed using rDNA technology was used to evaluate the level of federal regulatory agency involvement, to identify strengths, weaknesses and areas of potential improvement. The review concluded that none of the previously approved products of rDNA technology has had any significant negative impact on the environment.

Although all the case studies were published, OSTP/CEQ failed to reach a consensus on issues relating to the relevant strengths and weaknesses of the existing regulatory structure within the time allotted for the completion of its review. A review of the case studies published provides a comprehensive interpretation of the responsibilities of each federal regulatory agency in ensuring the safety of the products developed using rDNA technology.

National Academy of Sciences (NAS)

The National Academy of Sciences (NAS), and its operating arm, the National Research Council (NRC), have served as a primary source of scientific, technological, human health and environmental policy advice during the development of regulatory approaches for the safety evaluation of products developed using rDNA technology within the United States.

In 1987, the NRC published a report concerning the potential human health and environmental hazards associated with the commercial introduction of GM organisms, entitled *Introduction of Recombinant DNA-Engineered Organisms into the Environment: Key Issues.* The risks associated with the introduction of GM organisms were considered to be essentially the same in kind as those associated with unmodified organisms.

In other words, rDNA technology did not appear to introduce any unique risks as compared to the products that had been developed using more conventional methods of genetic modification. In reaching these conclusions, the NRC performed an evaluation of the similarities and differences in the properties exhibited by products developed using a variety of different techniques. To this day, the conclusions of this report continue to be referenced by the regulators and developers of GM crop varieties worldwide.

A subsequent NRC report, entitled *Field Testing Genetically Modified Organisms: Framework for Decisions,* also reached similar conclusions, but provided additional guidance as to how regulatory decisions concerning the introduction of GM organisms should be made. The NRC recommended that regulatory decisions concerning the introduction of GM organisms should be made on a case-by-case basis.

Consistent with the Coordinated Framework, the NRC did not consider the nature of the process used for the genetic modification of an organism to be a useful criterion for determining whether a product requires less or more regulatory oversight. As a result, no valid reason existed to regulate organisms genetically modified via modern techniques (*e.g.* rDNA technology) any differently from organisms genetically modified via more conventional means. Similar conclusions have been published in the reports of international standards-setting organizations.

In retrospect, within both reports, the NRC acknowledged that modern and conventional methods of genetic modification are not without risks to human health or the environment, and as a result, neither could be considered inherently more risky. As a result, regulatory decisions concerning the safety of the products of rDNA technology need only to take into consideration the specific characteristics exhibited by a particular GM organism and the environment in which it is to be introduced, and not the method by which it has been produced. The NRC has further articulated their conclusions into what is now commonly referred to as the 'Concept of Familiarity'.

Although familiarity with the characteristics of a particular organism or the environment to which it will be introduced would not necessarily mean it was safe, it can be expected to provide a sufficient amount of information to allow for a judgement to be made of the risks. For example, familiarity with a new GM plant variety could be established based on comparisons between characteristics of the parent line or other crop species exhibiting similar traits, as well as through the results of actual field tests involving the GM plant. These principles were further elabourated upon by the Organisation for

Economic Co-operation and Development (OECD), and as a result, have been referenced in the development of regulatory policies for evaluating the safety of GM crops on a global basis.

More recently, a committee established by the NRC published a report entitled *Genetically Modified Pest-Protected Plants: Science and Regulation*, based on a review of all scientific and regulatory data collected during the regulatory approval process for GM crops within the United States. The primary objective was to assess independently the effectiveness of existing and proposed regulations for the safety evaluation of GM crops expressing plant pesticides (*i.e.* plant-incorporated protectants). No new evidence was identified to suggest plants expressing plant pesticides posed any greater risk to human health or the environment as a result of their genetic modification. In fact, the NRC concluded, 'with careful planning and appropriate regulatory oversight, commercial cultivation of GM plants is not expected to pose higher risks and may pose less risk than other commonly used chemical and biological pest-management techniques'.

However, the NRC report included requests for federal regulatory agencies to further strengthen the current regulatory approval process through better coordination and communication between agencies, on-going investment in the research and monitoring of potential human health (*e.g.* allergenicity) and environmental impacts (*e.g.* insect resistance), and by providing greater access to information evaluated in support of regulatory decisions.

US FEDERAL REGULATIONS FOR AGRICULTURAL BIOTECHNOLOGY

Regulatory systems for the products of agricultural biotechnology have been in existence since the mid- to late 1980s within the United States. The regulatory approach to the safety evaluation of plants developed using rDNA technology has evolved in the best interests of research scientists, industry and the general public. The agricultural products of rDNA technology, such as GM foods and crops, may require approvals from up to three regulatory agencies; the US FDA, the USDA and the US EPA, depending upon the characteristics exhibited by the GM plant, its proposed use and introduced traits. The same standards of safety are applied to all products regardless of the technology used in their development.

The US FDA is responsible for ensuring the human safety of all new foods and food components, including products developed using rDNA technology, under the Federal Food, Drug, and Cosmetic Act (FFDCA). The USDA evaluates the potential of a GM plant to become a plant pest following its environmental introduction under the Federal Plant Pest Act (FPPA). The US EPA evaluates pesticides, including plant systems modified to express pesticides (*e.g.* insect-protected or virus resistance), under the Federal Insecticide, Fungicide, and Rodenticide Act (FIFRA). As a result, the expression of an insecticidal protein in a food crop would undergo review by the USDA,

US EPA and US FDA; a GM food crop exhibiting a modified oil content would be evaluated by the USDA and US FDA; and a non-food horticultural plant developed using rDNA technology for any other purpose (*e.g.* flower colour) would be subject to review by the USDA alone.

In most instances, obtaining all necessary approvals for the commercialization of an agricultural crop developed using rDNA technology takes a decade or more. However, the exact amount of time required will depend on the need to confirm performance, to evaluate characteristics of the food, environmental effects, and to produce the required amount of seed before the product can be distributed and commercially grown by farmers. Up to five years of field trials (5-10 generations of plants) are required for the developer of a new plant variety to collect sufficient data to meet the reporting requirements of the USDA. An additional five months to two years may be required for the US FDA, USDA and/or US EPA to complete all necessary product consultations, reviews and approvals.

Approval for the first commercial planting of a GM food crop was not issued until 1995. Since 1995, more than 40 new agricultural crops developed using rDNA technology have received approval for commercial planting within the United States. In 1999, approximately 72 per cent of the total 39.9 million hectares (more than 98 million acres) of GM crops grown worldwide were planted in the United States. Herbicide-tolerant soybeans (54 per cent), Bt corn (19 per cent) and herbicide tolerant canola (9 per cent) accounted for approximately 82 per cent of the GM plants cultivated. With an increasing number of agricultural biotechnology products reaching later stages of commercial development, it is anticipated that the overall area planted with GM crops will continue to rise.

The regulatory approach to evaluating the human health and environmental safety of GM crops within the United States is best described as a science-based, case-by-case assessment of hazards and risks. This approach has provided the flexibility required to reduce the regulatory burden placed on products that have been determined to be of low risk or concern. All agencies involved in the regulation of plants developed using rDNA continue to implement and develop policies based on recommendations made within the Coordinated Framework. In the future, the US FDA, USDA and the US EPA will be dedicating additional resources towards communicating how GM food and food components are regulated within the United States, and how these regulations function to be protective of both human health and the environment.

FOOD AND DRUG ADMINISTRATION (US FDA)

Authority

The US FDA is responsible for ensuring the safety and wholesomeness of all food and food components, including the products of rDNA technology, under the FFDCA. The US FDA has the authority for the immediate removal of any product from the market that poses potential risk to public health or that is being sold without all necessary regulatory approvals. As a result, a legal burden is placed on developers and food manufacturers to ensure the commodities utilized and foods available to consumers are safe and in compliance with all legal requirements of the FFDCA. In order to understand the regulatory approach followed by the US FDA in the safety evaluation of GM crops, it is useful to consider food and food safety from a historical context.

People had been consuming foods derived from agricultural crops for many years prior to the existence of any food laws or regulations within the United States. Based on this experience, agricultural crops have been accepted as being safe for consumption as food, without additional testing to demonstrate their safety. As long as the new crop variety has exhibited similar agronomic properties, and an appropriate taste and appearance, it has been considered safe to consume. As a result, most foods consumed today, in particular whole foods (*i.e.* fruits, grains and vegetables) and conventional foods, have not been subject to any kind of premarket review or approval by the US FDA.

Nevertheless, food scientists have a good understanding that many of the commonly consumed agricultural crops contain natural toxicants (*e.g.* tomatine in tomatoes, solanine in potatoes, cucurbiticin in cucumber, psoralens in celery, etc.). As a result, new plant varieties may be subject to routine chemical analyses to ensure that none of these substances is present at potentially harmful levels. This type of general approach has been used in assessing the safety of thousands of new plant varieties that have been developed over a number of decades of crop breeding without compromising the safety of whole foods.

Role

Consistent with recommendations within the Coordinated Framework, the US FDA considered existing provisions of the FFDCA to be sufficient for the regulation of foods and food components developed using rDNA technology. It was concluded that the scientific and regulatory issues posed by the products of rDNA technology were not significantly different from those posed by conventional products. As a result, GM foods and food components have been subject to the same standards of safety as already exist for the regulation of other foods and food components under the FFDCA. In order to better communicate interpretations of existing provisions of the FFDCA as they relate to the safety evaluation of foods derived from new plant varieties, including the products of rDNA technology, the US FDA released a policy statement in 1992 entitled 'Statement of policy: foods derived from new plant varieties'.

The US FDA has considered the use of genetic modification (*i.e.* rDNA technology) in the development of new plant varieties to represent a continuum of conventional plant breeding practices (*e.g.* mutagenesis, hybridization, protoplast fusion, etc.), and as a result, the safety evaluation of all new plant varieties, not just those developed using rDNA technology, have been evaluated based on an objective analysis of the characteristics of a food or its components, and not on its method of production.

Definition and Scope of Bioengineered Foods

The recently proposed rule of the US FDA concerns 'bioengineered foods' which have been defined as 'foods derived from plant varieties that are developed using *in vitro* manipulations of DNA (generally referred to as rDNA technology)'. As a result, the proposed rule has a much narrower focus than the 1992 US FDA Statement of Policy.

The US FDA has explained the need for a change in emphasis based on their expectations that many of the new plant varieties exhibit a greater potential to 'contain substances that are significantly different from, or that are present in food at a significantly different level than before'. As a result, the substances present in foods and food components derived from new plants developed using rDNA technology are less likely to be considered GRAS, and as a result, will require pre-market approval from the US FDA.

Safety and Nutritional Evaluation

The primary objective of the safety and nutritional evaluation is to demonstrate that the food derived from a new plant variety is as safe or nutritious as foods already consumed as a part of the diet. For new plant varieties, including those developed using rDNA technology, a science-based approach is used to focus the evaluation on the demonstrated characteristics of the food or food component. The evaluation of a GM food or food component typically involves reviewing information or data on any newly introduced substances, the known levels of toxicants, as well as the nutritional composition of the plant following modification. Substances that raise safety concerns (*e.g.* toxicants, allergens) would be subject to more extensive evaluation, since both intended and unintended changes may affect the levels of toxicants and nutrients in a food following the modification.

Guidance for performing a safety and nutritional evaluation was provided in the 1992 US FDA Statement of Policy, in a series of flow charts and text that cover:

- The crop that has been modified;
- Source(s) of the introduced genetic material;
- New substances intentionally added to the food as a result of the genetic modification (*e.g.* proteins, but also fatty acids, and carbohydrates).

Documentation required to support the evaluation typically includes: the purpose or intended technical effect of the modification on the plant, together with a description of the various applications or uses, a molecular characterization of the modification including the identities, sources and functions of introduced genetic material; information on the expressed protein products encoded by introduced genes; information relating to the known or suspected allergenicity and toxicity of any expressed gene products; for foods known to cause allergy, information on whether the endogenous allergens have been altered by the genetic modification; information on the compositional and nutritional characteristics of the foods, including anti-nutrients; and in some instances, comparative results of feeding studies involving the foods derived from plants modified using rDNA technology and the non-modified counterpart.

In performing its evaluation, the US FDA is particularly interested in the identification of inherent toxicants, known or potential allergens, assessing the concentration and bioavailability of essential nutrients, the safety and nutritional value of any newly introduced proteins, and the identity, composition, and nutritional value of modified carbohydrates, fats and oils.

If additional questions of safety remain following this evaluation, further toxicological studies may need to be performed. It is recognized that absolute assurance of the safety of any food does not exist. As a result, the goal of the safety evaluation is to establish a reasonable certainty of no harm under anticipated conditions of consumption. With this in mind, experience with the existing food supply has provided the basis for evaluating the safety of new food or food components. Both the Food Advisory Committee and Committee for Veterinary Medicine have been extensively involved in the development of approaches for the safety and nutritional evaluation of foods and food components derived from new plant varieties, including those developed using rDNA technology.

PRODUCT CHARACTERIZATION

Product characterization takes into consideration information relating to the modified food crop, the introduced genetic material and its expression product, and acceptable levels of inherent plant toxicants and nutrients. All characteristics of the gene insert must be known, including the source(s), size, number of insertion sites, promoter regions, and marker sequences. It must be established that the transferred genetic material does not come from a pathogenic source, a known source of allergens, or a known toxicant-producing source. The introduced genetic material should be well characterized to ensure that the introduced gene sequences do not encode harmful substances and are stably inserted within the plant genome to minimize any potential opportunity for undesired genetic rearrangement.

Analytical data are required to evaluate the nutritional composition, the levels of any known toxicants, anti-nutritional and allergenic substances, and the safety-in-use of antibiotic resistance marker genes. Any new substances introduced into crops through rDNA technology (*e.g.* proteins, fatty acids, carbohydrates) will be subject to pre-market review as food additives by the US FDA, unless substantially similar to substances already safely consumed as a part of foods, or that are considered GRAS. To date, substances that have been added to foods through rDNA technology have been previously consumed or have been determined to be substantially similar to substances already consumed as a part of the diet. As such, introduced substances have been considered exempt from the requirement for pre-market approval as food additives with the US FDA.

A more rigorous safety evaluation of a GM crop is warranted if the introduced gene sequence(s) has not been fully characterized, the nutritional composition has been significantly altered, antibiotic resistance marker genes have been used during its development, or if an allergenic protein or toxicant has been detected at levels higher than what is typically observed in edible varieties of the same crop species. In any event, determinations as to the safety of substances that have been introduced into new plant varieties through rDNA technology are made on a case-by-case basis.

Although an evaluation of the introduced gene sequences and expression product(s) provides assurance as to their safety, further studies may be required to predict whether unexpected effects may result following their interaction with other genes within the plant. In addition, the product characterization of a GM plant involves assessing sequence homology to known toxicants and allergens, thermal and digestive stability, and if required, the results of both *in vitro* and *in vivo* assays to demonstrate lack of toxicity.

COMPOSITIONAL ANALYSIS

The results of field trials performed over several years serve to characterize the phenotypic and agronomic characteristics exhibited by the plant (*e.g.* height, colour, leaf orientation, susceptibility to disease, root strength, vigour, fruit or grain size, yield, etc.), as well as to provide the materials required for the compositional analysis.

Any anomalies in the phenotypic or agronomic characteristics exhibited by a plant may result in a requirement for additional information. Protein, fat, fibre, starch, amino acid, fatty acid, ash and sugar levels are determined, as well as the levels of anti-nutrients, natural toxicants or known allergens. Studies of the nutritional composition are performed to determine whether the levels of any key nutrients, vitamins or minerals have been altered as a result of the genetic modification.

Based on the results of these studies, a determination is made as to whether the phenotypic and agronomic characteristics of a GM crop or the concentrations of inherent constituents fall within ranges typical of its conventional counterpart.

If inserting a new gene causes no change in any of the assessed parameters, the US FDA can conclude with reasonable assurance that the GM crop is as safe as the conventional crop. If the levels of essential nutrients or inherent toxicants are found to be significantly different in the GM crop, the US FDA may recommend additional action prior to commercialization, such as obtaining food additive status, or the use of specific labels to alert consumers of an altered nutritional content, etc.

ALLERGENICITY

In consultation with scientific experts in the areas of food safety, food allergy, immunology, biotechnology and diagnostics, the US FDA published guidelines for assessing the allergenicity of GM foods or food components in 1994.

The approach to assessment is multi-faceted, incorporating data regarding the origin of the genetic material, and the biochemical, immunological and physicochemical properties of the expressed protein. The overall assessment is reliant upon the fact that all known food allergens are proteins and, notwithstanding the number of shared properties between allergenic and non-allergenic food proteins, food allergens tend to exhibit a number of similar characteristics. In general, food allergens share a number of common properties: they have a molecular weight of over 10 000 Da; they represent more than 1 per cent of the total protein content of the food; they demonstrate resistance to heat, acid treatment, proteolysis and digestion; and they are recognized by IgE.

For gene sequences derived from known allergenic sources (*e.g.* peanuts), the developers of GM plants are expected to demonstrate that allergenic proteins have not been introduced into the food. For assessment purposes, it is assumed that any genetic material derived from a known allergenic source will encode for an allergen. To demonstrate otherwise, the amino acid sequence of an expressed protein must be compared with that of known allergens using protein sequence databases. Furthermore, *in vitro* and/or *in vivo* immunologic analyses using the sera of allergic patients sensitive to the source of the genetic material may need to be performed to determine whether or not a potentially allergenic protein is being expressed in the GM food.

Some GM foods may be modified to express genes from a source that is not known to be allergenic when consumed. Under these circumstances, the US FDA follows a similar decision tree-based approach to determining the allergenic potential of the expression product. In assessing these proteins, should any amino acid sequence exhibit homology with a known allergen, the expressed protein would then be evaluated in immunologic tests using the sera of patients known to be allergic to the identified homologous protein. Regardless of the origin of the genetic material, physicochemical studies are performed *in vitro* to provide information concerning the expected stability of the expressed protein. All known food allergens tend to be resistant to digestive degradation, as demonstrated in simulated gastric fluid models, or to decomposition under conditions of food processing.

In making a determination regarding a GM food, it is the totality of the biochemical, immunological and physicochemical properties of the introduced protein that provides guidance as to the allergenic potential of such a protein being expressed in food. The level of protein expressed, produced and consumed as a part of the diet is also a primary indicator of the allergenic potential, since nearly all food allergens are known to be major proteins in their respective foods. If the results of any of these studies suggest an allergenic potential, the US FDA may recommend further scientific evaluation, require special labelling to alert sensitive consumers, or alternatively caution the developer about proceeding with the development of a particular GM food.

Most recently, a joint FAO/WHO Expert Consultation on Allergenicity of Foods Derived from Biotechnology recommended a revised decision tree. This decision-tree strategy was modified from the previous version FAO/WHO to include a revised definition of sequence homology for gene product comparison; a greater emphasis on serum testing, even with gene products without homology to known allergens and not derived from an allergenic source; and animal models to assess potential allergenicity, despite acknowledgement by the Expert Consultation that these models are currently under development and at present, not predictive of food allergies in humans.

More recently, the *ad-hoc* Open-Ended Working Group on Allergenicity established by the *ad-hoc* Intergovernmental Codex Task Force on Foods Derived from Biotechnology, considered the FAO/WHO strategy in drafting an approach to assessing the potential allergenicity of foods derived from rDNA plants.

The Codex Working Group recognized the absence of a definitive predicitve test for allergenicity in humans to a newly expressed protein and recommended an integrated, stepwise assessment strategy. The strategy recommended by the Working Group, and accepted by the Codex Task Force for inclusion in the draft forwarded for final adoption by the Codex Alimenarius Commission, is consistent with that of FAO/WHO. However, the Working Group suggested that some of the modifications included in FAO/WHO could contribute to the overall weight of evidence of any conclusion of potential allergenicity (*e.g.*, allergen specific serum depositories, animal models), pending development and validation.

ANTIBIOTIC RESISTANCE

The use of antibiotic-resistance genes as selectable markers has been common practice in the development of new plant varieties using rDNA technology. Concerns relate to the potential transfer of antibiotic-resistance genes from GM plants to pathogens in the environment or to the gut of humans consuming foods or food components.

Issues relating to the use of antibiotic resistance genes were identified in the 1992 US FDA Statement of Policy Statement as well as discussed in additional guidance entitled *Guidance for Industry: Use of Antibiotic Resistance Marker Genes in Transgenic Plants*. The guidance provided within these documents was established in consultation with experts in the fields of microbiology, medicine, food safety, bacterial and mycotic diseases, and includes suggestions with respect to the continued safe use of antibiotic-resistance marker genes by the developers of new plant varieties.

The use of marker genes that encode resistance to clinically important antibiotics has raised questions as to whether their presence in food could reduce the effectiveness of oral doses of the antibiotic or whether the gene present in the DNA could be transferred to pathogenic microbes, rendering them resistant to treatment with the antibiotic.

The risk of transfer of antibiotic-resistance genes from plants to microorganisms considered to be pathogenic to humans, however, is considered to be minimal if not insignificant. Furthermore, the potential risks are becoming less of a concern as more developers are beginning to research the use of alternative technologies (*e.g.* non-resistance-based markers) in plant breeding. The conclusions with respect to the safe use of antibiotic-resistance marker genes are consistent with the findings of other national and international food safety organizations.

Consultation and Filing Process

The submission of a Pre-market Biotechnology Notification (PBN) has recently been proposed as a mandatory requirement for the commercialization of bioengineered foods and food ingredients within the United States. Minimal differences exist between the information to be submitted as a part of a PBN and that previously presented in voluntary consultations with the US FDA. As proposed, developers are still being encouraged to consult with the US FDA as early and as often as necessary in the development of a bioengineered food, such that any potential scientific or regulatory concerns can be identified and addressed prior to the submission of the PBN.

Guidelines for performing consultations with the US FDA were released in a publication entitled *Guidance On Consultation Procedures: Foods Derived from New Plant Varieties* in 1997. The publication recommended an approach for developers to proceed by submitting a request for consultation, outlined the internal process by which all requests would be handled, and included additional guidance as to the type of safety and nutritional information to be presented during consultation with the US FDA.

Once sufficient safety and nutritional information has accumulated to demonstrate that a product is safe and in compliance with the FFDCA, developers typically schedule a consultation to present their scientific findings and conclusions to the US FDA. Consultations prior to notification not only serve to keep the US FDA informed of advances made in the application of rDNA technology in food production, but also keep the developers of bioengineered foods aware of emerging safety, nutritional or regulatory concerns of the US FDA.

Consultations are considered complete when all safety and regulatory concerns between the US FDA and the developer have been resolved. The US FDA proposes to perform an initial evaluation of a PBN within 15 days of receipt to determine completeness, at which point, if considered complete, the PBN will be filed, and a response can be expected within 120 days. The US FDA does not issue a product approval *per se*, but informs the developer by letter that:

- The evaluation period has been extended;
- The notice is not complete and why;
- It has no further questions 'at this time' based on the information that has been presented.

Labelling

The FFDCA defines what information must be disclosed to consumers on a food label, such as the common or usual name, and other limitations concerning the representations or claims that can be made or suggested about a food product. All foods must be labelled truthfully and not be misleading to consumers. Taking this into consideration, the FFDCA does not stipulate the disclosure of information on the basis of consumer desire to know. Labelling may be considered misleading if it fails to reveal material facts in light of representations that are made with respect to a product.

The labelling of foods derived from new plant varieties, including plants developed using rDNA technology, was originally addressed in the 1992 US FDA Statement of Policy, and most recently discussed in Draft Guidance for Industry for the voluntary labelling of bioengineered foods.

To date, the US FDA is not aware of any information that would distinguish foods developed using rDNA technology (*e.g.* bioengineered foods) as a class from foods developed through other methods of conventional plant breeding, and as such, have not considered the method of development a material fact requiring disclosure on product labels. Nevertheless, after extensive consultation, including thousands of written comments and a series of public meetings, the US FDA has observed 'a general agreement that providing more information to consumers about bioengineered foods would be useful'.

Requirements for Labelling

Special labelling is required if the composition of the bioengineered food differs significantly from its conventional counterpart. For example, for a food that has been genetically modified to contain a new major sweetener, a new common or usual name or other labelling may be required. Similarly, if a GM food contains an allergen that consumers would not expect to be present in that food, special labelling may be necessary to alert sensitive consumers.

If a protein commonly associated with an allergic reaction (*e.g.* peanut protein) is transferred to another food through genetic modification, the US FDA would evaluate whether labelling would provide sufficient consumer protection. If labelling would not be considered to provide a sufficient level of protection, the US FDA would take appropriate steps to ensure the GM food would not be marketed. Therefore, current policy requires a GM food to be labelled when the resulting product poses a safety issue or is substantially different from its conventional counterpart, and as a result, could be considered to pose a misrepresentation to consumers.

The 1992 US FDA Statement of Policy and the recent Draft Guidelines do not consider the use of rDNA technology in the development of food products to be a material fact requiring specific disclosure on the label. Rather it is a method of development, similar to other methods of plant breeding, which have not required disclosure on the label. Bioengineered foods cannot be distinguished compositionally from foods modified through more conventional methods and thus do not require specific disclosure through labelling. The Draft Guidelines reaffirm the US FDA position that bioengineered foods do not require special labelling.

Voluntary Labelling

To provide guiding principles for voluntary labelling, in recognition of the desire of certain manufacturers to label foods as produced either with or without bioengineering, the US FDA published a 'Draft Guidance for Industry: voluntary labelling indicating whether the foods have or have not been developed using bioengineering'. Emphasizing that the use of rDNA technology 'is not a material fact', the US FDA recognizes that some consumers want disclosure of bioengineered content and that some manufacturers wish to provide it. In response, Draft Guidance was issued with suggestions concerning the use of labelling statements that are not considered misleading.

PRESENCE OR USE OF RDNA

The US FDA provided several examples of how disclosure of bioengineering can be accomplished, be informative and not be misleading.

- *Example* 1 'Genetically engineered' or 'This product contains corn meal that was produced using biotechnology'. These disclosures reveal the minimum amount of optional information about bioengineering.
- *Example* 2 'This product contains high oleic acid soybean oil from soybeans developed using biotechnology to decrease the amount of saturated fat.' This statement explains the proper and required name of this type of soybean oil, since it differs from what would be considered as soybean oil. The optional comments about biotechnology and decreasing saturated fat provide information that could be seen as benefits of the product.
- *Example* 3 'These tomatoes were genetically engineered to improve texture.' This example was one included to illustrate how it could be misleading to consumers if they cannot discern a difference in texture, but would not be misleading if they can tell a difference. If the former, and the new texture is to facilitate processing, then this intention should be made clear, *i.e.* 'to improve texture for processing'.
- *Example* 4 'Some of our growers plant tomato seeds that were developed through biotechnology to increase crop yield.' This is

another example of optional information that would explain an indirect, agricultural benefit.

ABSENCE OR NOT BIOENGINEERED

The US FDA provides an important commentary in the Draft Guidance about 'genetically modified organisms (GMO)' versus 'bioengineered'. It would be technically inaccurate to use the phrase 'not genetically modified' or 'GMO-free' to mean that bioengineering was not used. This is because most conventional foods have been genetically modified over the years by traditional crop breeding practices. Examples of acceptable voluntary statements would include:

- We do not use ingredients that were produced by biotechnology.'
- This oil is made from soybeans that were not genetically engineered.'
- Our tomato growers do not plant seeds developed using biotechnology.'

Another important point is made about a term such as 'GMO-free' that is misleading for two reasons:

- Many foods do not contain organisms anyway and therefore should not be labelled as 'organism-free';
- Free' implies complete absence or 'zero' amount of bioengineered material.

In a practical sense, it is impossible to demonstrate analytically the complete absence of anything. Therefore, in reality, a threshold for possible adventitious presence of a low level of bioengineered material may be necessary and has been the subject of much debate.

The Agency also provides additional guidance about misleading statements that:

- Could be interpreted to suggest that the absence of bioengineering would make the food superior to the bioengineered alternative;
- Claim the absence of one bioengineered ingredient when the food contains another ingredient that is bioengineered;
- Claim that a food is not bioengineered when in fact this type of food (*e.g.* green beans) has never been modified through rDNA technology.

DEPARTMENT OF AGRICULTURE

The Animal and Plant Health Inspection Service (APHIS) of the USDA is responsible for the protection of domestic agricultural resources and the environment from the threats posed by pests and diseases. The Federal Plant Pest Act (FPPA) has been the primary statute adapted by APHIS in regulating the products of agricultural biotechnology.

Under the FPPA, APHIS has the authority to restrict the import of plant pests, to introduce measures (*e.g.* quarantines) to prevent their movement, or to destroy plant pests cultivated in violation of the FPPA. The requirements under the FPPA are equally applicable to scientists within company, academic, research, and private organizations.

Definition of Plant Pests

A plant pest is any agent, substance or organism, or parts thereof, that may cause injury, disease or damage, either to plants or to the environment in which it is introduced. APHIS maintains a list of organisms considered to be plant pests, and therefore, subject to regulation under the FPPA. If an organism is not on the list, it may still be subject to regulation as a plant pest if there is reason to believe it is or will act as a plant pest. Organisms considered plant pests, or that exhibit the potential to be plant pests, are treated as 'regulated articles' under the FPPA.

The regulations extend to the introduction of GM organisms meeting the definition of a plant pest or for which there is reason to believe they are plant pests. APHIS considers the environmental release of a GM crop to be equivalent to the introduction of new organisms. Therefore, until proven otherwise, a GM crop is considered a 'regulated article' under the FPPA. The use of the term 'plant pest' in reference to GM plants means the 'non-pest' nature of the crop has yet to be determined. To date, all field tests conducted have demonstrated that GM crops exhibit no more 'pest-like' traits than their conventional counterparts.

All GM plants developed to date have involved the use of at least one designated 'plant pest' as either promoters or vectors, and as a result, have been subject to regulatory review under the FPPA. APHIS takes the position that an entire plant can be designated as a regulated article, even if it was not developed using plant pests, if the plant is the product of genetic engineering, and the agency determines or has reason to believe it is a plant pest. Given this position, APHIS would likely challenge any attempt to introduce a GM plant into commerce unless the developer has first gone through the APHIS regulatory review process.

Requirements for Commercialization

A number of years of field testing are required to evaluate the agronomic and product quality characteristics exhibited by new plant varieties developed within a labouratory or greenhouse. APHIS regulations outline procedures for obtaining a permit or providing notification prior to the importation, interstate movement or release of a regulated article, which of course, includes field testing. Prior to an introduction, a proponent must provide notification to APHIS of its intentions or submit an application to obtain a permit. The notification and permit requirements have evolved as an extension of the long-standing programm for the regulation of plant pests within the United States. Since 1987, APHIS has issued permits or acknowledged the receipt of notifications for the field release of more than 6500 new crop varieties.

Permits

For APHIS approval, developers must demonstrate that a new plant variety poses no significant risk to other plants in the environment or is at least as safe as similar crop varieties. APHIS review and approval are required for the shipment of seed and for the conduct of field trials involving new crop varieties. APHIS issues site-specific permits for field tests or releases into the environment.

Prior to field testing, APHIS may seek further clarification of study objectives, specify how, when or where research may be conducted, establish the data to be generated, collected and reported, stipulate additional requirements for the monitoring or securing of test sites, and specify methods for the disposal of crop residues that remain at the end of a study. The results of field trials are often used in making a determination as to whether a GM plant exhibits any effects on non-target species, or poses any unique plant pest problems.

For field trials conducted under either notification or permit, developers of GM plant varieties are required to submit reports to APHIS that include: details regarding the methods of observation; the data collected; and interpretations concerning the effects on plants, non-target organisms, or the environment. In the event that a GM crop is likely to be the subject of a petition for a determination of non-regulated status, developers must provide a description of known and potential differences, and substantiate that the regulated article is not likely to pose a greater plant risk than the organism from which it was derived.

An environmental assessment is a requirement of the NEPA, Council on Environmental Quality regulations, and USDA procedures for issuing permits under the FPPA. Under the permit process, developers of GM plant varieties must disclose information concerning the plant, test facilities, and control measures in place for its transport and field testing. Based on this information, the USDA performs an assessment of the potential environmental impact following a release. If the agency reaches a 'Finding of No Significant Impact' (FONSI), a permit may then be issued.

Notifications

Depending on the plant species and intended introduction, applying for a permit from APHIS can be a complicated and time-consuming process. In the spring of 1993, APHIS introduced a simplified notification process as an alternative to applying for a permit, which in most instances applies to the introduction of GM plants. To qualify for notification, a GM plant must be introduced in accordance with the eligibility criteria.

Originally, six GM crops (corn, cotton, potatoes, soybean, tobacco and tomatoes) qualified for notification; however, the process was more recently expanded to include most other crop species not capable of becoming a noxious weed. The developers of new GM plant varieties are encouraged to contact APHIS when making a determination as to whether a GM plant would qualify for notification. If a particular GM plant does not qualify for notification, a more involved regular permit process must be followed. For a GM plant to qualify for notification, it and its predecessor must not be a noxious weed, the introduced gene sequence(s) must be stably integrated within the host, not originate from a plant or animal pathogen, exhibit a known function that will contribute to plant disease, or encode for substances that could be potentially toxic or infective to non-target organisms, such as pharmaceuticals, or viral components other than those coat proteins already established as safe.

Performance standards also exist for the conduct of field trials to ensure that an adequate level of containment is provided for all introductions. It is the responsibility of the developer to ensure that an adequate level of containment is provided for field trials involving GM plants under notification. The performance standards utilized will vary according to the biology of the plant and the nature of the introduction. In general, this includes specifying methods for the proper handling, shipment, field monitoring and disposal of plant material at the end of the study.

As a part of the notification process, an applicant must provide information about the plant, identify the source of any genes used; the method of genetic modification; and the size, date and location of any proposed field test or introduction. Notification is required at least 10 days prior to the interstate movement or 30 days in advance of the field testing or importation of a regulated article that qualifies for notification with APHIS.

Determination of Non-regulated Status

Once sufficient data have accumulated through labouratory and field trials performed under permit or notification, a developer may petition APHIS for a determination of non-regulated status. A DONRS allows a GM plant to be grown, tested or used for crop breeding without further regulatory oversight by APHIS. APHIS will grant a DONRS if sufficient evidence can be provided to demonstrate that no plant pest potential exists (*i.e.* that the GM plant is not expected to become a pest, poses no significant risk to the environment, or is as safe as conventional plant varieties).

The developer is required to demonstrate a lack of change in disease or pest resistance status, an absence of any potential for contributing to development of a new plant pathogen or pest, as well as to address any questions relating to potential environmental consequences of the release.

APHIS examines several parameters when making a determination as to whether a GM crop should be considered a plant pest. Reporting requirements include submitting a detailed rationale for development; overview of the biology of the crop (competitiveness, survivability, dormancy); molecular characterization; protein expression; morphologi-cal/phenotypic characteristics; outcrossing/geneflow; weediness; insect and disease susceptibility; and, recombination potential (for viral genes).

Other considerations include assessing whether a GM crop could itself become a weed. In making this determination, factors that are considered include: how the seeds are dispersed in the environment; whether the seeds are capable of surviving over the winter; the potential for the development of volunteer plants; and whether these plants could reproduce into viable offspring. Another important consideration is whether the seeds or plants are capable of surviving outside of a managed agricultural environment (*e.g.* watering, fertilizer, etc.), and how the introduced traits could influence the viability of the crop under marginal conditions.

APHIS also takes into consideration potential adverse effects on wildlife, including birds, beneficial insects and mammals, following the introduction of a GM crop. Some of this information comes from field trials that are conducted in multiple locations for several years. By observing crops growing under actual conditions of use in the field, scientists can compare insect populations existing within a field planted with GM crops to those coexisting in fields planted with a non-modified variety of the same plant. Field trials are also useful in identifying any changes in the plant physiology, such as height, colour, leaf placement, time of flowering, etc., that result from modifications in the genome of the plant.

Monitoring of these changes during field trials allows APHIS to evaluate how these changes might benefit or adversely impact the behaviour of wildlife in contact with these crops. Knowing that wildlife, such as deer, often feed on agricultural crops, APHIS also takes into consideration the nutritional content of the GM crop. Comparisons of the essential nutrient content are often made between the GM plant and its conventional counterpart. Other interests may include determining the impact of the modification on the levels of adverse factors (*e.g.* natural toxicants or anti-nutrients) present in the plant.

APHIS maintains a list of all crops that have received a DONRS. At any time, APHIS retains the authority to prohibit the commercial planting of a GM crop if it is determined that it is becoming a plant pest.

PETITION PROCESS

Before a GM plant can be freely transported and commercialized, the developer must first petition the APHIS for a DONRS. A petition for a DONRS requires the submission of extensive data on the introduced gene construct, effects on plant biology and effects on the ecosystem, including the spread of the gene to other crops or wild relatives. Developing all data necessary to support a petition for a DONRS may take months or years depending upon the scientific issues that need to be addressed.

Based on the information provided within the petition, APHIS evaluates the potential impact (*e.g.* toxicological) that the introduced modifications will have on non-target organisms in the environment, including threatened or endangered species. A petition for a DONRS must provide a rationale for the development and introduction of the GM crop, starting with genotypic and phenotypic information on both the host and donor organism(s). This includes a detailed description of the methods of transformation, identifying all gene sequences (*e.g.* promoters, leader sequences, introns, selectable markers, etc.), their function, and potential for plant risk following introduction. A combination of different analytical methods may be used to demonstrate the stable integration of an introduced gene sequence within the genome, its inheritance and expression in host plants. To the extent possible, the gene(s) of interest (*e.g.* resistance, marker genes), levels of expression, and im pacts on levels of inherent plant toxicants in plant tissues, under experimental and actual growth conditions in the field, must be characterized.

Field performance studies (*e.g.* leaf morphology, pollen viability, seed germination, viability, insect susceptibilities, disease resistance, yield, etc.) are used to identify any differences or similarities in phenotypic characteristics exhibited by the GM and non-modified crop. The extent of field performance data required in support of a petition for non-regulated status depends not only on the nature of introduced gene sequences, but also on the physiology of the host plant. Less extensive field performance studies are required for plants that are: highly domesticated (*e.g.* corn); self-pollinating (*e.g.* soybean); male sterile; those with high seed germination rates (> 90 per cent); or unlikely to influence their potential weediness or fitness (*e.g.* delayed ripening, oil seed content). Whereas, GM crops exhibiting a tolerance or resistance to cold, salt, biotic (*e.g.* insects, pathogenic agents), or other abiotic (*e.g.* herbicide) stresses, are subject to more extensive field performance testing.

APHIS considers several parameters in reaching a DONRS, including whether the GM plant can cross-pollinate with other plants in the wild, and if it can, the ecological consequences (*e.g.* insect resistance, herbicide tolerance) that might result. Extensive databases are maintained on species capable of cross-pollinating with GM crops (*e.g.* corn, soybean, cotton, canola, etc.), based on the results of years of breeding experiments, biological surveys and research conducted by agricultural and weed specialists within APHIS. In instances where out-crossing might be expected to occur, APHIS may stipulate the planting of refuge areas to limit the potential for adverse ecological effects following commercial planting.

AGENCY RESPONSE

APHIS will grant a determination of non-regulated status if the developer can sufficiently demonstrate a lack of plant risk. In making a determination that a GM plant will no longer be regulated, APHIS prepares two documents, an environmental assessment and a DONRS to satisfy regulatory requirements under NEPA and FPPA, respectively. Both documents are developed based on the review of data submitted as a part of the petition for a DONRS. Once the DONRS has been granted by APHIS, a GM crop no longer requires notification or permit prior to its movement or release within the United States.

A review for DONRS is generally completed within 10 months of submitting a formal application to the APHIS, allowing sufficient time for publication in the Federal Register, and opportunity for public comment.

Notices of all petitions for a DONRS are published in the Federal Register, and the public is given 60 days to comment for or against the petition, after which APHIS has up to an additional 180 days to either approve or deny it. APHIS provides access to a considerable amount of information relating to approved field trials and petitions received for GM crops. The petition for a DONRS must include scientific details regarding the genetics of the plant, the nature and origin of the genetic material used, the potential for indirect effects on other plants, and any other information that could be considered unfavourable to the petition.

Extensions of Non-regulated Status

In certain instances, the non-regulated status of a GM crop may be extended to include other varieties of the same crop species, provided that changes in introduced gene sequences are insignificant, and no new plant pest issues would be expected. For an extension, developers must substantiate that the new regulated article poses no serious issues meriting review under separate petition, by providing a precise description of the genetic modifications to the regulated article and a detailed comparison to modifications made in the GM plant previously granted non-regulated status by APHIS. Field trials must be performed under notification or permit to demonstrate the similarities in the phenotypic properties. In support of multiple extensions, product identity standards may be submitted describing the genotypic and phenotypic properties exhibited by new plant varieties considered extensions of an existing DONRS for a GM crop.

Future Considerations

The USDA recently appointed an Advisory Committee for Agricultural Biotechnology that will work in conjunction with the NAS (Standing Committee on Biotechnology Food and Fibre Production, and the Environment) on the completion of a critical review of the regulations and policies used by APHIS in the commercial approval of GM crops. The committee comprises scientists, farmers and representatives of consumer groups and seed companies, and is anticipated to take two years to complete its assignment.

In addition to APHIS, other functional areas within the USDA are beginning to get more actively involved with the regulation of agricultural biotechnology. For example, the Grain Inspection Packers and Stockyards Administration (GIPSA) has made a commitment to work with farmers and industry groups in the development and validation of reliable test methods and quality assurance programms to differentiate GM from non-GM commodities. As a part of this commitment, GIPSA intends to publish a proposed rule on the standardization of test methods and approaches for the detection and identity preservation of grains. GIPSA is also intending to offer accreditation services for labouratories testing grains for the presence of GM content, and to evaluate commercial test kits to ensure they can be considered accurate and reliable. It is expected that an administrative fee will apply to the accreditation services being offered by GIPSA.

The USDA also has made continuous commitments to funding research that will assist federal regulatory agencies in making science-based decisions concerning the safe introduction of GM organisms into the environment. Active areas of research include assessing the potential risks associated with the environmental introduction of GM plants (*e.g.* gene flow), the cumulative effects of large-scale commercial plantings, the potential interactions between GM and non-GM crop species, programmed resistance, and the development of statistical methodology and quantitative approaches to measuring the risks associated with the field testing of GM plants, etc.

Environmental Protection Agency

The EPA has authority over the registration of chemical and biological pesticides under the Federal Insecticide, Fungicide and Rodenticide Act (FIFRA). A pesticide is any product 'intended for preventing, destroying, repelling, or mitigating any pest'.

All pesticides require registration by the US EPA prior to their distribution. As a part of the registration process, an applicant is required to submit information and data in support of the safety of a pesticide in its intended use. In addition, US EPA is responsible for establishing acceptable tolerance levels for registered pesticides on or in raw agricultural food commodities under the FFDCA. At any point, US EPA has authority to amend or revoke a registration or residue tolerance, if an adverse effect is observed, or if the risks associated with the use of the pesticide are determined to be unacceptable.

Role

Under the Coordinated Framework, the US EPA was identified as being responsible for the review, assessment and registration of the products of rDNA technology that act as pesticides. Leading up to this publication, the US EPA had acquired considerable expertise in the registration of biological pesticides (*e.g.* microbial). The US EPA was first involved in discussions regarding the potential risks of pesticides developed using rDNA technology in the early 1980s, during a series of public meetings involving the FIFRA Scientific Advisory Panel (SAP) and the Biotechnology Scientific Advisory Committee (BSAC).

It was recognized that the potential risks posed by plants modified to express pesticidal components through rDNA technology were unique and warranted additional consideration relative to conventional pesticides. For example, the potential for introduced genetic material to out-cross from a modified plant to a sexually compatible wild or weedy relative through pollen spread was one of the unique, theoretical risks posed by plant-pesticides.

As a result, the EPA proposed a plant-pesticide rule in 1994, outlining their interests in the regulation of products. A package of three proposed final rules was released early in 2001, redefining regulations for plant-incorporated protectants (PIPS), previously referred to as plant-pesticides. Essential elements of the proposed rules for the regulation of GM plants expressing pesticidal traits have been followed since 1994. The final rules provide further clarification of how PIPS will be regulated under FIFRA and FFDCA, including specific exemptions and became effective in September 2001.

Definition of Plant-incorporated Protectants (PIPS) or plant-pesticides

US EPA's final rule clarifies the regulation of whole plants that contain PIPS. When plants are used intentionally for controlling pests in some way, such uses meet the definition of a pesticide under FIFRA. Nevertheless, the agency exempts the plants themselves from regulation, focussing instead on the substances they contain, previously referred to as plant-pesticides and now known as PIPS. A plant-incorporated protectant (PIP) includes the substance produced by the living plant, and all genetic material necessary for its production (including promoters, enhancers, etc.).

Thus, genes and substances (*e.g.* enzymes) responsible for the conversion of natural plant constituents into pesticidal substances would be considered PIPS. To date, PIPS registered by the US EPA have been proteins and the genes required to make these proteins within the plant. Crops involved include potatoes, cotton, field corn, sweet corn and popcorn.

Clearly, Bt toxins represent the most predominant class of PIPS that have been registered by the US EPA. Major reasons for this include the success of microbial products containing Bt and their long history of use for nearly 40 years. It has been estimated that combined, Bt corn, potato and cotton were cultivated on approximately 10 million acres in 1997, 20 million in 1998, and 29 million in 1999. As a result, less chemical insecticide has been applied, and there have been significantly higher crop yields. In addition, due to reduced opportunity for opportunistic fungal infections of insect-damaged corn, lower levels of mycotoxins have been produced, reducing the toxicological risk to both humans.

Exemptions

US EPA has concluded that living plants with pesticidal activity do not require a high level of regulatory scrutiny by the agency and thus, are exempt from regulation under FIFRA. The agency commented that focussing on the substance produced within the plant, *i.e.* the PIP eliminates any need for the US EPA to register plants, thus conserving agency resources and reducing potential overlap with other agencies, notably the USDA. For example, although whole plants that are used as biological control agents themselves, such as chrysanthemums, are exempt, substances that are extracted from plants, such as the insecticidal material pyrethrum, are not excluded from regulation under FIFRA. Further exemptions have been proposed, based on familiarity and presence of the pesticidal substances in the food supply. Only one of the additional proposed exemptions has been included in the final rule, and all others have been proposed for further public comment. Within the final rule, EPA has exempted PIPS and encoding genetic material originating from plants sexually compatible through conventional breeding but not through rDNA technology. The rationale for this exemption is that such substances could be bred, either naturally or through conventional plant breeding, from close plant relatives that share genetic material from a common gene pool.

On the other hand, genetic modifications that were never before possible can be made using rDNA technology, involving the genes from sexually incompatible plant species. Therefore, the exemption of plant-pesticides developed using rDNA technology, involving genes from sexually compatible plants has not been finalized as a part of the supplementary proposal to the final rules for PIPS.

Other exemptions for further comment have also been identified within the supplemental proposal. The first group are PIPS that 'act primarily by affecting the plant'. The group have in common some sort of defence mechanism such that a pest would be less able to attach to, penetrate, or successfully invade a plant. Examples include:

- Some sort of structural barrier such as wax or plant hairs;
- Inactivation of or resistance to a pest or substance (*e.g.* toxin) produced by a pest;
- Creation of a deficiency in a nutrient or growth factor required by a pest;
- Hypersensitivity response that would contain or limit spread of the pest (*e.g.* necrosis of surrounding plant tissue, creating a functional barrier);
- Plant hormones that are naturally occurring but with change can affect a plant in a variety of ways with potentially dramatic results.

The other classes of PIPS proposed for exemption include viral proteins and their encoding genes. The expression of genes for viral proteins within a host plant confers resistance to infection by that virus as well as certain related viruses. These PIPS are plant virus specific, and as such, would not be expected to pose any risk to the public

Table. Viral Proteins

Protein	Virus
Viral coat protein	Cucumber mosaic
	Papaya ringspot
	Potato Y
	Watermelon mosaic
	Zucchini yellow mosaic
Replicase	Potato leaf roll
Modified from NRC.	

Herbicide-tolerant Crops

In the case of herbicide-tolerant crops, EPA determines whether the property of increased resistance may encourage higher rates or more extensive application of an agricultural chemical to food crops. The higher exposures to herbicides would be subject to evaluation through risk assessment, requiring the submission and review of detailed data. If the risk is determined to be acceptable, a herbicide product would require label extensions for the addition of new tolerant crop varieties.

Experimental Use Permit (EUP)

Small-scale research involving PIPS is covered under the permit and notification process that exists for new plant varieties administered by APHIS. The US EPA requires that an Experimental Use Permit (EUP) be obtained for all research and field testing involving greater than 10 acres of a plant modified to express a pesticidal substance. The EUP process allows a developer to gather the product use, performance, residue and other types of data, required in support of the registration of PIPS. All US EPA decisions concerning applications for EUPs are published within the Federal Register for comment.

Registration requirements

For the registration of a plant-incorporated protectant, the EPA evaluates the potential impact, on human health and the environment, of the pesticide substance and of the genetic material necessary for its expression within the modified plant. With this in mind, the EPA developed their guidance for evaluating the safety of PIPS based on data requirements that were established for microbial pesticides.

A number of modifications to the reporting requirements were required to focus the assessment on non-target toxicity rather than pathogenicity, as well as to evaluate the expression of the pesticide substance within the plant itself, rather than by spraying, etc. Although specific reporting requirements for the registration of PIPS are established on a case-by-case basis by the EPA, the general categories of information and data required in support of the registration of a plant-pesticide include:

- Product characterization
- Toxicology
- Environmental fate and exposure
- Non-target organism effects
- If necessary, plans for pest resistance management.

The EPA continues to hold meetings of their SAP to re-evaluate the reporting requirements that have been established for PIPS and to ensure that all potential human health and environmental concerns are being addressed as new scientific information becomes available. In addition, the developers of PIPS are encouraged to consult with US EPA during all phases of the research and development process, particularly when it concerns an environmental release. As a part of their review, scientific evaluators within the EPA weigh the potential hazards inherent to the generic consideration of PIPS and the potential for exposure to the protectant substance. The characterization of possible hazards, in the context of exposure, allows for estimates of potential environmental risk to be established.

The commercial approval of a PIP may take between 15 and 18 months following receipt of a full data package by the US EPA, and it involves a public notification and consultation process. However, this time frame would be considered an ideal, since in reality, deciding what constitutes a 'full' data package may take several rounds of discussions with the US EPA.

Table. Environmental Risk Assessment of Plant-incorporated Protectant

Potential hazard	Exposure potential
Inherent nature of gene(s)	Feeding on plant or pollen
Effect on host plant relatives	Gene transfer to other plant
Effect on non-target susceptible species (*e.g.* plants; invertebrate and vertebrate animals; microbes)	Concentration in plant part of gene product Degradation/persistence of gene product
Disruption of ecosystem checks and balances	Release/movement of gene product
Weediness	Production of pollen Distribution of seed

Product Characterization

Product characterization allows for consideration of any potential hazard(s) that might be associated with anticipated exposure to a plant-incorporated protectant. This process includes a review of the source of introduced genetic material, including all regulatory sequences, the modifications that have been made, methods of transformation, inheritance, stability and expression in the host plant. For the introduced pesticidal substance, the anticipated modes of action, specificity and toxicity to both target and non-target organisms are evaluated. Target organisms may include weeds, insects and diseases (viral, bacterial, and fungal) affecting a crop. The crop itself, as a volunteer weed, may also be of concern following crop rotations. Once a PIP and any potential for exposure have been characterized, the data required in support of an environmental and human health risk assessment can be defined.

Ecological Effects

Environmental fate and exposure are important considerations when evaluating the potential impact of PIPS on the environment or on non-target organisms (*e.g.* wildlife, beneficial insects). Depending on the crop and introduced pesticide substance, data on the level of expression and rate of degradation may be required, first for the tissues of the modified plant, and secondly, the fate in soil, water, or any other environmental compartment. Such data allow the potential for exposure to the pesticide substance to be evaluated for non-target organisms via feeding on, or through decomposition of, the modified plants.

Studies are arranged in tiers, starting with Tier I, consisting of acute toxicology and basic characterization studies that help predict environmental fate. Depending on the results obtained in Tier I, and the level of concern, additional studies may be requested from higher tiers. It is expected that most PIPS will only undergo Tier I testing, since negative results are expected at relatively high exposure levels. The 'maximum hazard approach' when used in Tier I provides a high level of assurance that adverse effects would not be expected to occur under actual conditions of use within the field. If the results of Tier I reveal a potential hazard at doses within the range of expected concentrations in the environment, additional studies would be requested in higher tiers to better characterize the potential for any kind of effect.

Data requirements may be satisfied by generating test data or by requests for data waivers with credible justification. The type and extent of dietary studies required will depend on properties of the crop and on the introduced pesticide substance. Dietary studies typically involve feeding plant tissues (*e.g.* grain, pollen) that express the PIP or that have been spiked with the pesticidal substance.

However, when determined necessary, the pure pesticidal substance may be administered to non-target species at doses of up to 10 to 100 times the expected level of exposure in the field. Dependent upon the potential for exposure, additional studies may also be requested involving any number of other non-target organisms (*e.g.* honeybees, green lacewing, ladybird beetles, parasitic wasps, earthworms, etc.). When toxicity is observed, the potential for exposure becomes an important consideration when ascertaining whether an adverse effect might be expected under actual conditions of use in the field.

The requirement for provision of a resistance management plan is not routine practise for PIPS, but depends on an evaluation of the potential for resistance to develop and the threat to existing or functionally equivalent pest management practises. To date, resistance management plans have only been placed on Bt PIPS. If considered appropriate, the EPA may request that a resistance management plan be submitted as a part of a condition for registration. A resistance management plan may require a commitment towards the generation of additional research data, to implement structured refuges, to perform annual resistance monitoring, to initiate remedial action plans, and to educate growers. The EPA has worked closely with academia, other federal agencies, public interest groups, industry, and growers, to establish resistance management plans based on the most current science.

Human Health Effects

For agricultural food crops, special consideration is given to evaluating the safety of the pesticidal substance introduced as a component of consumed foods. Rodents (often mice, to conserve test material) are orally administered the pesticide substance at doses of up to 100 000 times the levels of expected human consumption as a part of the diet.

Although the modified plant tissues would be considered the most appropriate vehicle for determining the acute toxicity of a PIP, it is not typically feasible, since only small amounts of the pesticidal substance are present in plant materials. Therefore, an alternative source (*e.g.* microbial) is often used to generate a sufficient amount of identical pesticide substance for testing purposes. In this case, the equivalency of the test substance obtained from an alternative source to the PIP (*e.g.* composition, identity, etc.) being expressed within the plant is an important consideration that must be well documented.

For pesticide substances that are proteins, the principal concern relates to the potential for the introduction of a new allergen or toxin into the food supply. Since most allergenic food proteins are stable to digestion, *in vitro* digestibility studies are used to predict how long a pesticidal substance may persist in digestive fluids (*e.g.* gastric, intestinal) following ingestion. In addition, heat stability studies are performed to provide an indication of the fate of the PIP during food processing.

Commonly, assays of amino acid sequence homology are performed to determine similarity to substances that are known to pose unique safety concerns (*e.g.* toxins, allergens). The EPA uses a 'decision tree' approach similar to that followed by the FDA in assessing the potential allergenicity of a new food protein. In many instances, US EPA and US FDA collabourate with each other when evaluating issues of safety for new pesticidal proteins being expressed in GM food crops.

Residue Tolerances for Plant-incorporated Protectants

Analogous to the consideration of tolerances for residues of conventional pesticides, the US EPA reviews data on the human, animal and environmental safety of PIPS to determine whether residue tolerances should be established for the amounts expected to be present in foods derived from the GM plant. Under Section 408 of the FFDCA, if the GM plant expressing a PIP is a food crop, the US EPA is obligated to establish the 'safe level' of pesticide residue allowed, or tolerance level.

The Food Quality Protection Act (FQPA) served to amend the FFDCA and FIFRA by outlining the process by which the safety of pesticide residues on raw or processed foods can be established. Under this process, in order for a pesticide residue to be considered safe, a 'reasonable certainty of no harm' must exist assuming aggregate exposure of the public, taking into consideration sensitive children within the population. To date, all pesticidal proteins that have been registered as PIPS have been determined to be non-toxic, and as a result, have been considered exempt from the requirement to establish a tolerance under the FFDCA.

Consistent with categorical exemptions as originally proposed for plant-pesticides under FIFRA, the US EPA also proposed that tolerances might not be required for pesticidal substances derived from sexually compatible plants, regardless of the process used to introduce the PIP, provided that the genetic material encoding the pesticidal substance was from a plant commonly used as food, and that the presence of the pesticidal substance in a host plant did not result in any new or significantly different dietary exposures. Final rules include exemptions from tolerances for the residues of PIPS, and their encoding nucleic acids, but only for PIPS derived through the conventional breeding of sexually compatible plants. US EPA is requesting further comment on the possible exemption from the requirement of a tolerance for:

- PIPS derived through rDNA technology;
- PIPS that act by affecting plant defences;
- PIPS developed based on viral coat proteins.

Future Considerations

The US EPA anticipates that the requirements for the registration of PIPS will continue to evolve as the science and policies relating to biotechnology continue to mature. In light of on-going criticisms that the plant-pesticide rule has received since its publication in 1994, a number of stakeholder workshops have been hosted by the US EPA. In March 1999, the House Agricultural Subcommittee hosted a congressional hearing to address a number of these concerns. One of the primary questions addressed was whether the definition of a pesticide under FIFRA provides the US EPA with the authority to regulate pest-protected plants that have been developed using rDNA technology.

Another concern was that the proposed rule was too broad in scope because all plants have natural defence mechanisms, and as such, would be classified as 'plant-pesticides'. In addition, an exemption from the plant-pesticide rule for those plants not developed using rDNA technology would serve to regulate the process used in the development of the GM plant, rather than on the pesticidal substance being produced within the plant. Further criticism related to the use of the term 'plant-pesticide' in relation to the products of the genes that had been introduced into GM crops.

As a part of its final rule, the US EPA recommended that plant-pesticides be renamed PIPS, an alternative name which distinguishes them from the other types of pesticide products requiring registration with the US EPA. A new FIFRA section has been added to the Code of Federal Regulations specific to the regulation of PIPS. In doing so, the US EPA has acknowledged that the new final rule would be making a distinction between the products of conventional breeding and rDNA technology, *i.e.* regulating on the basis of process not product. One reason given for this divergence was that the use of this criterion would be expected to provide the public with increased confidence that an appropriate level of regulatory oversight was in place for assessing the risks of rDNA technology, which is a societal not scientific issue.

SAPs are charged with providing the Office of Pesticide Programms (OPP) of the US EPA with advice and guidance on human health and environmental issues relating to proposed regulatory decisions for pesticides, including PIPS, based on the latest scientific data. For example, the SAP for Bt Plant Pesticides met three times in the year 2000 to consider issues related to the safety of plants that had been genetically modified to express insect-specific Bt toxins.

The US EPA released a decision to temporarily extend registrations for all currently registered pest protected plants throughout the 2001 growing season. Additional provisions, such as the recently strengthened resistance management requirements for plant-pesticides, along with the original registration conditions, were considered to provide adequate protection during the extended time period. In addition, the reassessment process for these products has been designed to assure maximum transparency and opportunity for public comment in the regulatory decision-making process.

DISEASE RESISTANT CROPS

CUCURBITS AND SQUASH

Squash belongs to the Cucurbitaceae, a family of crops that is an important source of food throughout the world. While not as important as the cereals or oilseed crops, such as soybean, nor other vegetable crops such as potatoes and tomatoes, cucurbits are nevertheless grown on significant acreages worldwide.

In 1998, cantaloupes and other melons were grown on 1 044 672 hectares, producing 17 764 188 metric tonnes of food; cucumbers and gherkins were grown on 1 567 389 hectares, producing 26 673 943 metric tonnes; while pumpkins, squash, and gourds were grown on 1 136 083 hectares, producing 14 169 983 metric tonnes. Cucurbit crops, including squash, are used for food in all regions of the world, but are also used for oil and other non-food uses, such as musical instruments or household implements.

IMPACT OF VIRUS DISEASE ON SQUASH

Squash, like the other major cucurbit crops, is susceptible to infection by several viruses. The most damaging are cucumber mosaic virus (CMV), papaya ringspot virus watermelon strain (PRSV-W), watermelon mosaic virus 2 (WMV2) and zucchini yellow mosaic virus (ZYMV). While the extent to which viruses cause economic damage in cucurbits, and squash in particular, is not well documented, the problem is extensive enough that diligent effort to breed squash resistant to CMV, PRSV-W, WMV2 and ZYMV has been a goal of breeders for many decades.

DISEASE BREEDING IN SQUASH

Sources of genes for resistance or tolerance to viruses have been reported in various relatives of squash and in *Cucurbita pepo* itself. However, introducing useful levels of resistance to more than one virus into commercially acceptable cultivars has been difficult to achieve. While releases of zucchini-type and English marrow-type varieties carrying resistance to a single virus (CMV) has been achieved, breeding for multiple virus resistance has only recently been successful. A summer squash cultivar, Whitaker, has recently been developed by researchers at Cornell University.

This cultivar is resistant to multiple pathogens, including viruses and fungi. The difficulty of this accomplishment is emphasized by the fact that Whitaker was the result of work conducted by a team of researchers for more than a decade. Because the sources of resistance were related species not normally sexually compatible with *C. pepo*, these researchers had to employ a commonly used breeding tool, embryo rescue, to overcome the barriers to crossing between species.

SQUASH AS A SPECIES

Cucurbita pepo is a crop native to North America. It consists of several subspecies and varieties that are completely interfertile, and are therefore presently considered part of a single species. *C. pepo* ssp. *pepo* includes cultivated pumpkins, marrows and a few ornamental gourds; *C. pepo* ssp. *ovifera* var. *ovifera* includes cultivated crookneck, scallop and acorn squashes, and most ornamental gourd cultivars; *C. pepo* ssp. *ovifera* var. *texana* includes free-living populations found in Texas; *C. pepo* ssp. *ovifera* var. *ozarkana* includes free-living populations found in the Mississippi Valley and the Ozark Plateau; *C. pepo* ssp. *fraterna* includes non-cultivated populations found in northeastern Mexico. Populations of *C. pepo* exist today under non-cultivated conditions in the United States.

These free-living populations range from northeastern Mexico and Texas, to Alabama, and through the Mississippi Valley to Illinois. Some free-living populations, especially in Arkansas, Louisiana and Mississippi, occur as weeds in soybean and cotton fields, and are difficult to control. Gene exchange between free-living and cultivated populations has been documented.

Furthermore, some free-living populations may have originated as escapes from cultivation, which might have undergone subsequent introgression with other nearby cultivated, or other free-living populations.

GENETIC ENGINEERING OF SQUASH

Because of the need to obtain useful virus resistance in a commercially acceptable cultivar, an alternative to breeding was attempted: the direct introduction of genes conferring resistance into already known commercial genotypes. The genetic engineering approach was to introduce the coat protein genes of the viruses in question, a strategy first successful in tobacco, and subsequently shown to be effective in cucurbit species.

Constructs containing viral coat protein genes were introduced into proprietary inbred lines of *C. pepo* ssp. *ovifera* var. *ovifera* via *Agrobacterium*-mediated transformation. The engineering of the constructs used in the transformations, as well as the transformation procedure, are described in Tricoli *et al.*.

Based on greenhouse testing and small-scale field tests, two lead lines, ZW20 and CZW3, were chosen as parents of commercial hybrids. The characteristics of these lines are described in detail by Tricoli *et al.*

ZW20

Line ZW20 was transformed with disarmed *Agrobacterium* harboring a binary Ti-plasmid. This plasmid contained expression cassettes consisting of the coat protein genes of zucchini yellow mosaic virus (ZYMV) and watermelon mosaic virus 2 (WMV2) under the control of the cauliflower mosaic virus 35S promoter. The T-DNA also contained, as a selectable marker, a neo-mycin phosphotransferase II (NPTII) gene from *E. coli* transposon Tn5, under the control of the cauliflower mosaic virus 35S promoter. The transformation event contained multiple insertion events, some of which were structurally complex. Among these insertion events were two loci containing genes encoding both the ZYMV and WMV2 coat proteins, but not the NPTII gene.

One locus provided a moderate level of resistance to the two target viruses, while the other provided a high level of resistance. Advanced breeding lines were made homozygous for the high-resistance locus, but were segregating for the moderate-resistance locus. During this process, the NPTII marker was also eliminated by selection of lines that not only were homozygous for the high-resistance locus, but had also lost the selectable marker through segregation.

CZW3

Line CZW3 was transformed with the plasmid pPRBN-CMV73/ZYMV72/WMBN22, containing expression cassettes for the coat protein genes for ZYMV, WMV2 and cucumber mosaic virus (CMV), as well as the NPTII selectable marker. Line CZW3 was a simpler transformant, consisting of only one insertion event, containing all three coat protein genes, as well as the selectable marker. Advanced breeding lines were homozygous for the transformed locus.

Field Results

Field trials at a number of locations showed that the lines ZW20 and CZW3 were effective in controlling infections by ZYMV, WMV2, and - in the case of CZW3 - CMV as well. Tricoli *et al.* mechanically inoculated two different ZW20 sublines, including one homozygous for the high-resistance locus. This subline showed only 20 per cent infection after inoculation with WMV2, and no infection at all after ZYMV infection, compared with 100 per cent infection of non-transgenic controls. The same researchers studied the resistance of CZW3 after mechanical inoculation with CMV, ZYMV and WMV2 individually, and in a mixed infection. Segregating inbred lines were tested, as well as a hybrid in which the segregating inbred was used as a parent. After individual and mixed inoculation of transgenic segregants, only one plant out of a total of 146 showed mild symptoms, while all of the 103 non-transgenic segregants as well as all of the 259 non-transgenic control plants became severely symptomatic. Further tests were carried out in the field, comparing virus resistance in these lines after mechanical and aphid inoculation. Other tests studied the economic benefit of the resistance provided by lines ZW20 and CZW3 either as an inbred line or as the parent to a hybrid. These tests confirmed that virus resistance conferred by the coat protein genes was able to dramatically increase crop yield and economic return under the pressure of naturally vectored disease in the field.

USDA EXEMPTION

The determination of efficacy against viral disease under field conditions led to the preparation of a petition for a determination of non-regulated status from the United States Department of Agriculture, Animal and Plant Health Inspection Service (USDA-APHIS). USDA-APHIS is the primary agency responsible for regulating work, in the greenhouse and field of transgenic plants. It also has the authority to grant non-regulated status to transgenic plants that in its opinion no longer present a plant pest risk. The main responsibility for conducting an assessment of safety rests with USDA-APHIS.

The assessment relies upon a review of the information package supplied to the agency by the petitioner, but the agency can and does seek out information on its own, and relies on the expertise of the reviewers on its staff. All relevant information is used by the agency to conduct an environmental assessment, and - if the assessment allows - grant non-regulated status. One guiding principle followed by the USDA-APHIS in assessing risk is to compare the characteristics of the transgenic plant with the characteristics of plants that are produced by traditional plant breeding. Plants produced by traditional breeding are not regulated, because history provides sufficient assurance of safety.

Thus, if the transgenic plant were determined to pose a risk different from that posed by plants already existing in nature or plants that would be produced by traditional breeding, then non-regulated status would be withheld. If, on the other hand, the risks presented by the transgenic plant are not different from those that are presented by traditionally bred varieties, then non-regulated status can be granted. Since USDA-APHIS relies heavily (but not exclusively) on the information contained in the petition for non-regulated status, information in such a petition should include the following:

- A description of the biology of the plant prior to the transformation. The biological description should include detailed taxonomic information.

- Relevant experimental data and publications relating to the work conducted on the transgenic plant.
- A description of the differences in genotype between the transgenic plant and the corresponding non-transgenic plant. Information must be provided on the organisms from which the transgenes or other control elements were obtained, the method of transformation and the vectors used during the transformation process, as well as the inserted genetic material and its product(s). USDA-APHIS, the Canadian Food Inspection Agency, and Health Canada, have standardized the types of molecular and expression data submitted in a petition for non-regulated status and the criteria by which they should be judged.
- A description of the phenotype of the transgenic plant, including any known and potential differences from the corresponding non-transgenic plant, in order to substantiate the claim that the transgenic plant is unlikely to pose a greater plant pest risk than the non-transgenic plant from which it was derived. This information should include the following: plant pest risk characteristics, disease and pest susceptibilities, expression of the gene product, new enzymes, or changes to plant metabolism, weediness of the transgenic plant, impact on the weediness of any other plant which is sexually compatible with the transgenic plant, agricultural or cultivation practices, effects of the transgenic plant on non-target organisms, indirect plant pest effects on other agricultural products, transfer of genetic information to organisms with which it cannot interbreed, and any other information which might be relevant. Any information indicating that a transgenic plant may pose a greater plant pest risk than the corresponding non-transgenic plant must be included.
- Field test reports for all trials of the transgenic line in question that were conducted under USDA-APHIS permit or notification. The field test reports should include any observations of deleterious effects on the environment caused by the transgenic plant.

Based on information received by USDA-APHIS from various sources including the petition, the Agency granted non-regulated status to ZW20 and CZW3, and published their support for this decision. In keeping with their goal of considering all information that might be available to them, USDA-APHIS made use not only of data supplied by the petitioners, but also relied on the literature, external expert opinion, and comparison with traits that were already familiar to agriculture. This approach acknowledged that scientifically valid risk assessment could be conducted by relying on information obtained from a wide variety of sources.

WEEDINESS

Of particular concern in determining the plant pest risk of transgenic squash was the potential for the crop itself to become a weed, or for new weeds to arise as a consequence of gene flow from the crop to sexually compatible wild populations. This question was especially important because the transgenic squash would be released in a region that was known to be the centre of origin for *C. pepo*. After evaluating all the information available, USDA-APHIS concluded that ZW20 and CZW3 were unlikely to cause a weed problem or a threat to genetic diversity of free-living squash if released from regulation.

With regard to causing a weed problem, the agency concluded that the transgenic lines were unlikely to become weeds themselves, and that the virus-resistance traits were unlikely to cause free-living squash to become weeds. The conclusion that the transgenic lines themselves were unlikely to become weeds was based on three reasons. First, the inbred lines used in the transformations belonged to the yellow crookneck squash type (*C. pepo* ssp. *ovifera* var. *ovifera*), which did not exhibit weedy characteristics and was not considered to be a weed in standard texts and weed lists. Second, introduction of CMV (for CZW3) as well as ZYMV and WMV2 resistance into the crop was not different from other disease-or pest-resistance traits already present in the *C. pepo* gene pool or introduced into it by traditional breeding.

These traits had already been introduced into squash, and even into some commercial cultivars, yet these cultivars had not become weeds. Finally, the two transgenic lines did not exhibit any other characteristics that were more 'weedy' than those exhibited by traditionally bred squash cultivars. Aside from considering the potential problem caused by the transgenic lines themselves, USDA-APHIS also evaluated the possibility that traits transferred from ZW20 and CZW3 to free-living relatives would cause those free-living plants to become more 'weedy'.

USDA-APHIS concluded that this problem was unlikely to happen. The agency assumed that the transgenic lines and their cultivated progeny would be able to pollinate free-living squash plants and transmit the virus resistance traits to them. This undoubtedly had occurred in the past with other disease- or pest-resistance traits that had been introduced into *C. pepo*. However, despite this opportunity, no weedy hybrid progeny had emerged. Since there was no evidence to suggest that the transgenic traits would behave differently from the traditionally bred traits, the historical lack of weed problems provided a sufficient level of assurance that crosses between transgenic and free-living squash were unlikely to create a weed problem as well.

Furthermore, the cultivated squash genotypes that were used for transformation possessed characteristics that placed those plants at a selective disadvantage in a non-cultivated habitat. The transmission of these traits to free-living relatives along with the transgene would place recipient plants at a disadvantage, further reducing their weedy potential. Finally, surveys of free-living squash populations taken by the applicant failed to detect infections of CMV, WMV2 or ZYMV. Therefore, there was no evidence that viruses limited the populations of free-living squash.

Consequently, it was not likely that virus resistance would release these populations from control, nor was it likely that virus resistance conferred a significant selective advantage in free-living habitats. The lack of selective pressure exerted by viruses was also cited by USDA-APHIS as the reason for concluding that genetic diversity in free-living populations would not be reduced. Without selection pressure, there was no reason to expect selection of a subset of genotypes into which the virus resistance would have been crossed.

TRANSCAPSIDATION/RECOMBINATION

The commercial introduction of viral sequences via transgenic plants - and in particular viral genes encoding viral coat proteins - present the potential for interactions between the transgenes or their products, and the viruses to which the transgenic squash might remain susceptible. These interactions include transencapsidation of heterologous viral RNA by coat proteins expressed in the plants, phenotypic mixing between viral coat protein subunits expressed by the transgenic plant or by the transgenic plant and infecting viruses, and recombination between transgenic viral components and other viruses infecting the transgenic plants. DeZoeten pointed out that these phenomena have the potential to produce new viruses. USDA-APHIS concluded that the risk of transencapsidation and phenotypic mixing in transgenic virus-resistant plants would be no greater than their likelihood of occurrence in non-transgenic plants already in agriculture. Data submitted by the petitioners, and other information gathered by the agency, showed that the amounts of viral coat protein produced in transgenic plants resistant to CMV, ZYMV and WMV2 were less than the amounts found in susceptible, virus-infected plants.

With respect to this set of data, USDA-APHIS noted that ZYMV and WMV2 coat protein levels increase when ZW20 plants were inoculated with ZYMV and WMV2 singly or combined. The agency speculated that this increase might be due to limited replication of the viruses in inoculated ZW20 plants, or because the coat proteins produced by the transgenes were stabilized by limited replication of the inoculated viruses.

EPA EXEMPTION FROM TOLERANCE

As is the case with USDA-APHIS, the US Environmental Protection Agency (EPA) also has the primary responsibility of conducting a safety assessment, based on information supplied by the developer of a transgenic crop and other information that the agency might obtain, including that of external and internal experts. When the first transgenic squash line (ZW20) was ready for commercialization, the EPA did not have a final rule covering transgenic plants. However, consultation with the EPA took place during the time the agency was formulating its proposed rules.

These rules extended the agency's jurisdiction to all plants expressing disease-or pest-resistance traits, since they could be viewed as producing a substance that was pesticidal - that is, they mitigated the effect of a pest. While the agency claimed jurisdiction over these plants, it recognized that certain broad categories of plants could be exempted from regulation. For example, plants produced through traditional breeding, or plants that repelled pests or disease through the modification of some structural feature could be exempted since they usually presented no new pesticidal exposure to humans consuming those plants. Furthermore, the EPA proposed to exempt coat protein produced in transgenic plants from regulation under the Federal Insecticide, Fungicide, and Rodenticide Act (FIFRA) and the Federal Food, Drug, and Cosmetic Act (FFDCA), again because they presented no new exposures to humans consuming them. Even though the rules were only proposed, consultation with the EPA led to the decision to submit a petition for exemption from tolerance (the legal limit for a pesticide chemical residue in or on a food), as required by the FFDCA prior to the commercialization of any pesticide.

In order to obtain this exemption, it was important to show that the establishment of a tolerance level was not required to protect the public health. This goal was achieved for line ZW20 by measuring the amount of viral coat protein in samples of various melon and squash fruit on sale at major grocery stores in one representative locality in the United States. Results showed that levels of viral coat protein in transgenic fruit were within the range found for most of the fruit in the sample. For some samples, the fruit on the market had many-fold higher levels of coat protein than in the transgenic squash.

These data were used to support the claim that viral coat proteins in general, and ZYMV and WMV2 coat proteins in particular, were not substances that fitted the two main criteria which the EPA planned to use to determine whether a pesticidal substance would require review. First, in order to trigger EPA review the pesticidal substance would be one that was not derived from a known food source. Second, the substance would present the following changes in pesticidal exposure to humans: it would be found in significantly higher levels in the transgenic plant than that normally found in food, it was being transferred to a different crop than that in which it was already found, it was not already found in edible tissue, and it was not already found in unprocessed food.

Other arguments advanced in the petition in support of the exemption from the requirement of a tolerance were based on the premise that a demonstrated history of safe use was sufficient to assess safety. Examples from the literature showed that humans were already exposed to viral coat proteins in general, not only through consumption of naturally infected produce, but also through ingestion of food obtained from crops that had been subjected to cross protection (the intentional inoculation of a crop with a mild strain of a virus in order to protect the crop from subsequent infection by a more virulent strain of the virus).

The petition also argued that supported the claim that expression of ZYMV and WMV2 in ZW20 would not present any changed exposure of humans to allergens. Because ZYMV and WMV2 coat proteins are commonly present in cucurbit fruit, anyone who might be allergic to viral coat protein in squash (a phenomenon for which no documented cases were known) would have the same reaction to the transgenic fruit as to non-transgenic fruit they had already consumed. The petition also addressed exposures due to dermal contact and inhalation. The petition argued that a history of safety had been established through the lack of reports of toxicity or allergenicity from researchers who had undoubtedly been exposed to these viruses by both the dermal and inhalation routes.

Finally, the petition pointed out that viral coat proteins were ubiquitous in the environment, and therefore a natural component of it. Infected cucurbit fruit that are not marketable are routinely plowed back into the soil, thus adding large amounts of viral coat protein to the environment. Furthermore, studies in the literature have reported the recovery of plant viruses from ground water run-off in Europe and North America. The EPA considered the data and the arguments presented in the petition and granted an exemption from tolerance for the ZYMV and WMV2 coat proteins produced within ZW20 plants.

Subsequently, the EPA granted a more categorical exemption from tolerance for ZYMV and WMV2 coat proteins in all food commodities. The decisions and supporting reasoning are summarized in the following paragraphs, which are based on the published decision. A detailed review of this decision is especially instructive because it provides insight into the issues that by law must now be considered by the EPA. A detailed review also provides another example of the reliance of US regulatory agencies not only on data submitted by petitioners, but also on the literature and other sources of information. In granting the broader exemption from tolerance, the EPA considered several issues that by law they were obligated to address, especially as a result of the passage of the Food Quality Protection Act of 1996 (FPQA), which amended the FFDCA.

The amended FFDCA allowed EPA to establish an exemption from the requirement for a tolerance only if EPA determines that the tolerance is 'safe'. By law, 'safe' meant that there was 'a reasonable certainty that no harm will result from aggregate exposure to the pesticide chemical residue, including all anticipated dietary exposures and all other exposures for which there is reliable information'. This includes exposure through drinking water and in residential settings, but does not include exposure due to its use during application or other occupational activities. In deciding on exemptions from tolerance the EPA is also required to pay special attention to exposure of infants and children. The EPA examined two areas of concern in determining the risks from aggregate exposure to pesticide residues.

First, the agency examined the toxicity of the 'pesticides' (ZYMV and WMV2 coat proteins). Second, it examined exposure to the pesticide through food, drinking water, and through other exposures that occur as a result of pesticide use in residential settings. In considering toxicity, petitioners are required by law to submit data regarding acute toxicity, genotoxicity, reproductive and developmental toxicity, as well as subchronic and chronic toxicity. However the petitioner requested, and EPA granted, waivers from these data requirements. The waivers were based on the long history of mammalian consumption of the entire plant virus particle in foods, without causing any deleterious human health effects.

The EPA once again acknowledged that virus-infected plants currently are - and have always been - a part of both the human and domestic animal food supply, and that there is no evidence that plant viruses are toxic to humans and other vertebrates. Furthermore, the EPA pointed out that plant viruses are unable to replicate in mammals or other vertebrates, thereby eliminating the possibility of human infection. As additional assurance of safety, only the coat protein would be expressed in the plant, and this component alone would be incapable of forming infectious particles. The EPA also had to consider the toxicity of the introduced DNA (coding and regulatory regions) directing the synthesis of the coat proteins. The EPA acknowledged that DNA is common to all forms of plant and animal life and that it was consumed as a safe component of food.

The specific nucleic acids that were used to transform the squash plants were characterized by the petitioners, and a review of the nucleic acid sequence led EPA to conclude that no mammalian toxicity was anticipated from the genes encoding the coat proteins of ZYMV and WMV2. In considering exposures, the EPA had to consider available information concerning exposures from the coat protein in food and all other non-occupational exposures, including drinking water from groundwater or surface water, and exposure through its use in gardens, lawns or buildings. With respect to dietary exposure in food, the EPA concluded, as they did before, that the use of viral coat protein in squash would not result in any new dietary exposure to plant viruses, because:

- Entire infectious particles of ZYMV and WMV2 are found in the fruit, leaves and stems of most plants,
- Viruses are ubiquitous in the agricultural environment at levels higher than will be present in transgenic plants,
- Virus-infected food plants have historically been a part of the human (including that of infants and children) and domestic animal food supply, with no known adverse effects to human or animal health.

Moreover, the EPA concluded that there was no evidence to indicate that harmful effects occurred as a result of exposure to viral coat proteins through other means, such as non-food oral, dermal and inhalation routes. They arrived at this conclusion by relying not only on data submitted by the petitioners, but on the expert opinion of its Scientific Advisory Panel, which concluded that the levels of virus in the agricultural environment are much higher than those present in transgenic plants, and that the existing presence of viral coat proteins in the food supply provided a scientific basis for exempting transgenic plants that express viral coat proteins from the requirement of a tolerance.

With respect to drinking water exposure, the EPA concluded that potential exposures through this route would be negligible.

The agency reasoned that viral coat proteins produced in plants are an integral part of the living tissue of the plant, and are therefore subject to rapid degradation and decay. Consequently, viral coat proteins would not accumulate in the environment or in the food chain. Therefore, because of their presumed transience, these proteins would present only negligible exposure in drinking water obtained from surface or groundwater sources. The EPA used similar reasoning to conclude that exposure of engineered coat proteins as a consequence of their use in such areas as around homes, parks, recreation areas, athletic fields and golf courses, would be negligible. The EPA concluded that the likelihood of exposure to plant viral coat proteins, including ZYMV and WMV2 coat proteins, through any dermal or respiratory routes would be unlikely.

Although physical contact with the plant or raw agricultural food obtained from the plant may present some opportunity for dermal exposure, the potential amounts involved in this exposure were judged to be negligible compared with exposure through the diet. Furthermore, the EPA reasoned that the skin would provide a barrier that would not be crossed by viral coat proteins. Similarly, exposure through inhalation was judged to be negligible compared with potential exposure through the diet. The most likely opportunity for respiratory exposure through inhalation was through pollen. However, the EPA concluded that it was again unlikely that the amount of exposure to the coat proteins in the pollen would be significant. Viral coat proteins, when present in pollen, would be part of the pollen grain and would not cross the barrier presented by the mucous membrane of the respiratory tract.

Therefore, this route would not add any additional exposure to that presented by the dietary route. A new requirement imposed by the Food Quality Protection Act involved the assessment of available information, indicating that there might be special concerns regarding exposure of infants and children to viral coat proteins. The EPA concluded that viral coat proteins were present in all foods, including those foods consumed by infants and children. It had no evidence to indicate that the amount of exposure was different between adults and children, nor that the effect of viral coat proteins differed between the two groups.

Most importantly, the EPA had no evidence suggesting that exposure of adults or infants and children was in any way harmful. Finally, the EPA had to consider whether ZYMV and WMV2 coat proteins, or the DNA needed to produce them in plants, would have an effect on the immune and endocrine systems. Because the agency had no information indicating that such an effect could be expected, it did not require that the petitioner submit information regarding that question. The decision to grant an exemption from tolerance for ZYMV and WMV2 coat proteins in ZW20 and subsequently in all raw commodities also cleared the way for commercialization of CZW3. EPA then granted an exemption from tolerance for the coat protein of CMV in all raw commodities, based on the same safety assessment it conducted for the other two coat proteins. Line CZW was included in this exemption and was therefore cleared for commercialization.

FDA FOOD SAFETY ANALYSIS

In contrast to USDA-APHIS and EPA, the United States Food and Drug Administration (FDA) makes the safety assessment of new plant varieties, and the foods made from them, the responsibility of the developer or manufacturer. In 1992, the FDA published guidelines to clarify the obligations of any producer of new plant varieties, whether produced through traditional plant breeding, or through techniques of recombinant DNA. The principle governing the guidelines is that assessment of the safety of a product (in this case food produced from a new plant variety) should focus on the product's characteristics rather than the method by which it was made.

The authority of FDA to regulate food is a post-market authority. Thus, a food producer can market a food containing a new component that it has assessed to be generally recognized as safe (GRAS). If this component is not GRAS, then a food additive petition must be filed and granted before food containing that new component can be marketed. The 1992 guidelines were meant to help producers in determining whether a food safety concern existed. If a concern existed, a consultation with FDA would take place to determine whether a new component could be considered GRAS, or whether a food additive petition was required.

In practice, consultation with FDA regarding new transgenic plants that will become food usually occurred regardless of safety concerns. The consultation assures producers that FDA is aware of the developer's safety assessment process, and serves to identify those parts of the safety analysis that might not be acceptable to the Agency. This process enables the producer to proceed towards commercialization with confidence that no unanticipated safety concerns will be encountered at late stages of product development. Recently, the FDA has published its intention to make this consultation process mandatory.

The assessment applies equally well to CZW3. In general, the safety assessment specified by the guidelines examines the host plant (the plant which has been modified by introduction of foreign DNA), the organisms from which the foreign DNA has been obtained, and the new compounds produced as a result of the introduction of the foreign DNA.

First, the guidelines direct the developer to determine whether there is a history of safe consumption of the species that was transformed. In the case of squash, the species has had a history of safe consumption for at least 3000 years, and is consumed in many forms throughout the world. The particular genotypes transformed to produce ZW20 and CZW3 have been in use commercially for several years.

The guidelines also ask the developer to consider whether characteristics of the host species, related species, or the lines used in transformation warrant analytical or toxicological tests. If so, they also ask whether the results of the tests provide evidence that the toxicant levels in the new plant variety do not present a safety concern. Squash indeed has characteristics warranting analytical testing, since the plant is known to produce cucurbitacins, which are extremely bitter toxic alkaloids. Squash breeders are aware that cucurbitacins must be eliminated from commercial varieties. Therefore, standard breeding practices take into consideration the potential presence of this compound, especially if genes from wild or free-living relatives are crossed into cultivated squash.

The FDA guidelines indicate that the types of testing currently used to screen for known toxicants have been effective in managing the risk of such compounds, and are therefore appropriate to apply to the case of transgenic crops. For the cucurbitacins, tasting fruits for bitterness has been an effective test method for the detection of this compound. The documented sensitivity of the human sense of taste for this compound is as low as 1-10 parts per billion, depending upon the particular form of cucurbitacin. At 10 parts per billion, this detection level is 34 000 times less than the reported oral LD50 in mice. Therefore, tasting fruit is an appropriate analytical test for the presence of cucurbitacin, and the presence of these compounds can be detected long before toxic levels are reached.

The non-bitter taste of ZW20 and CZW3 indicates the absence of cucurbitacins in these lines, as they were in the original lines before transformation. This result CMV, ZYMV and WMV2 coat proteins are almost identical to the coat proteins (presumably edible) found in cucurbit fruit. Squash containing viruses is not known to be allergenic, nor is it known to be toxic. Intake of viral coat proteins from the fruit of transgenic plants will be comparable to the intake of these proteins in non-transgenic infected fruit. Finally, the amounts of viral coat protein expressed in transgenic ZW20 and CZW3 are too low to present a possibility of becoming a major dietary constituent. These data were presented to the FDA in order to inform the Agency of the justification for the judgement that no food additive petition was warranted for ZW20 and CZW3 (they were therefore GRAS). Since no objections to these analyses were raised by the Agency, commercialization proceeded.

The assessment of safety of transgenic crops continues to be a topic of scientific and public concern. A review of the issues that surround the debate over the safety of these crops reveals that the issues addressed seven years ago during the commercialization of virus-resistant squash - which was the first commercialized virus-resistant crop - are still unresolved.

The appropriate perspective on safety is one that was expressed well before transgenic crops were at issue:

A thing is safe if its risks are judged to be acceptable. By its preciseness and connotative power, this definition contrasts sharply with simplistic dictionary definitions that have 'safe' meaning something like 'free from risk.' Nothing can be absolutely free of risk ... Because nothing can be absolutely free of risk, nothing can be said to be absolutely safe. There are degrees of risk, and consequently there are degrees of safety.

This is the perspective that is taken by the agencies responsible for regulating transgenic crops in the US. They all judge the level of risk of these crops by comparing them to the existing level of risk, such as risks presented by crops developed by traditional breeding or risks presented by natural phenomena. Furthermore, even though the risk presented by these crops might be greater than those presented by current unregulated activities, the acceptability of that risk also determine whether these crops are safe enough to commercialize and use. It is this latter issue that adds the dimension of public perception and acceptance to the debate.

The deployment of transgenic crops will most likely continue and increase with time. The potential benefits to the consumer, the farmer, as well as to the companies developing these crops will serve as a potent force for the exploitation of this technology. As this continues, scientific enquiry into the potential risks should continue. These investigations will help to further clarify areas of real concern. By focusing attention on these areas, which will probably be very few, safe use of this technology will be further improved.

FOOD SAFETY ISSUES

Although the variety registration of existing crop plants has not resulted in adverse effects in humans, the conventional assessments carried out by plant breeders was not considered to be sufficient to ensure food safety of GM crop plants.

As a result, Europe issued a regulation on novel foods and novel food ingredients that sets the legal framework for the market introduction of genetically modified organisms (GMOs). This came into force. The Novel Foods Regulation establishes a system of mandatory pre-market approvals of novel foodstuffs, including GM plants. The accompanying EU guideline gave a clear indication of the types of data that are needed to form the basis of an assessment. In particular, four central safety issues have been set forth related to food and feed safety:

- The nutritional and toxicological consequences of inserted gene products;
- The potential of pleiotropic (unintended) effects in the host organism due to the insertion event;
- The allergenicity of expressed proteins and novel foodstuffs;
- The potential of gene transfer to human and animal gut flora.

This means that in transgenic insect-resistant Bt tomatoes encoding the C-terminal truncated Bt2 gene derived from a Bacillus thuringiensis (Bt) strain, IAb5, the novel protein has to be evaluated on its own merit, while the safety and wholesomeness of the 'remaining' transgenic tomato is assessed separately.

Whereas, for example, in antisense RNA exogalactanase tomatoes, which have improved rheologic characteristics due to downregulation of the endogenous exogalactanase activity, the safety assessment is primarily focused on the 'remaining' novel fruit.

In general, there is basic agreement on the safety issues to be addressed. One particular area of concern is the possibility of unexpected or unintended metabolic perturbations due to genetic modification that may alter, for instance, levels of nutrients and health-influencing components. The use of rDNA techniques does not necessarily result in fundamental changes in crop plants compared to the food produced by conventional breeding methods. However, it should be emphasized that a uniform international agreement for the evaluation of GMOs is urgently needed. For example, an agreement on how to establish in practical terms the similarities or differences between the 'remaining' GM crop plant and the traditionally bred plant seems to be much more difficult to achieve.

The description of advanced strategies for the evaluation of potential unintended effects in GM crop plants. Pleiotropic (unintended) alterations in agronomic traits or composition may arise from insertion mutagenesis or as a result of metabolic effects of the novel gene product(s). The challenge is to gain the ability to discriminate between metabolic alterations due to somaclonal variations, unintended effects or natural diversity.

Thereto, emphasis is given to a platform of technologies designed to identify changes in the molecular machinery of crop plants. As is the case for detection methods for GM crops and derived products, there are needs and criteria for sampling strategies, statistical models and manufacturing standard reference matehers recently.

ASSESSING THE FOOD SAFETY

Strategies for assessing the food safety and wholesomeness of novel foodstuffs are currently in the exploration phase. Above all there is a need for uniform and harmonized quantitative methods for the testing of potential unintended effects in whole foods, such as GM crop plants are. The safety of, for example, synthetic chemicals and food additives is established by assessing separate elements of the compound in question. Although very successful in case of single chemicals, the evaluation of a whole food product may not be a simple addition of individual assessments of the constituents.

Traditional testing protocols involving labouratory animal 90-day feeding trials with (whole) food products are far from ideal. This type of animal feeding study with complex food matrices is complicated by the likelihood of nutritional imbalances leading to dietary problems, confounding factors, an insensitivity for specific endpoints, and the impossibility of using large safety margins, if any. Therefore, in a global context several regulatory, research and industry-based bodies have proposed alternative concepts for the safety evaluation of genetically engineered foods and food ingredients.

In particular the Organisation for Economic Co-operation and Development (OECD), the Food and Agriculture Organization (FAO) and the World Health Organization (WHO) have developed concepts and principles for the safety evaluation of GMOs and derived products. Their reports concluded that risk assessments should be directed at demonstrating that a GM crop plant or derived food product is as safe as a traditional or previously authorized product.

Accordingly the EU Competent Authority incorporated these general principles into their Novel Foods Regulation and accompanying guidelines. In this context, novel foods and food ingredients that are considered to be 'no longer equivalent' to traditionally bred plants are subject to labelling as defined in the Council Regulation.

REGULATION THE EVALUATION OF THE SAFETY AND NUTRITIONAL VALUE OF A GM PLANT

The term 'substantial equivalence' appears in the EU Novel Foods Regulation as a major principle in the risk assessment. It emphasizes that, within the context of the Regulation the evaluation of the safety and nutritional value of a GM plant should be focused on a comparative analysis with conventionally bred products. OECD's Group of National Experts on Safety in Biotechnology enunciated this comparative approach towards the safety assessment of transgenic food plants as the concept of substantial equivalence. In their report, it is assumed that existing food organisms possess a long history of safe use (*i.e.* GRAS).

Consequently, the most practical approach to the determination of safety is to consider whether the GM plants are substantially equivalent to analogous food product(s), if such exist. Once substantial equivalence of a novel foodstuff has been established, it provides assurance of safety that is equal to or better than that of its comparators. Further clarification of the concept of substantial equivalence originates from a WHO sponsored workshop. It has been concluded that the establishment of substantial equivalence is not a safety assessment per se, but a dynamic, analytical exercise.

Thus, it may mean that analysis of, for instance, gene expression patterns, global changes in protein expression and/or differences in metabolic capabilities (the metabolome) of the novel product in comparison with conventionally bred products should be performed in order to examine equivalence. This strategy has worked satisfactorily for the assessment of the first-generation crop plants for which sufficient background knowledge is available.

However, relatively less experience has been gained with the safety and wholesomeness evaluation of novel food plants for which no substantial equivalence can be established. Even in the case of single gene modification, potential alterations in metabolic pathways are difficult to identify and, consequently, the implications of the genetic modification process on the metabolism of plants are poorly understood. Moreover, data on the mechanisms by which plants regulate, for instance, their response to environmental stress factors and other conditions are remarkably scanty.

CONCEPT OF SUBSTANTIAL EQUIVALENCE APPEARS IN PRACTICE TO LEAD TO VARIOUS INTERPRETATIONS

Application of the concept of substantial equivalence appears in practice to lead to various interpretations. Different requirements for risk assessment with regards to the potential of unintended effects exist as a result of genetic modification.

Actual approaches entail a consideration of the characteristics of host and donor organism, phenotypic properties and toxicological evaluation and include comparative determinations of any changes in critical (anti-)nutrients and key toxicants for the food source in question. In general, the analysis of an expanded spectrum of components is unnecessary, but should be considered if there is an indication from other traits that there may be an unintended secondary effect of genetic modification. However, compositional analyses based on single parameters using a list of crop-specific critical (anti-)nutrients and key toxicants as a minimum requirement of screening for possible variations in the content has its limitations. Limited data are typically available for the less important food components, as in less well-known crops most critical (anti-)nutrients and natural toxicants will be unknown.

Moreover, the toxicants to be assessed may be partly influenced by the cellular function of the product encoded by the inserted gene. Furthermore, the natural variation of critical comparators may mask the evidence of secondary effects due to genetic modification. Thus, there is no criterion that determines whether a difference between a GM plant and its natural counterpart is outside the literature values for the ranges of parameters that express natural diversity.

Relevant information on indicators of unexpected effects may not have been observed during the development of an edible transgenic crop species. In the future other compounds could become of importance, such as anti-oxidative, oestrogenic and anti-carcinogenic compounds. In determining critical nutrients and/or key toxicants, differences in consumption patterns and practices of processing and consumption in various geographical settings and cultures must also be recognized. It is our opinion that conclusions about relative safety and wholesomeness based on a list of selected single components may not be equally valid in a global context and in all times. Therefore, there is an urgent need to set up generic approaches using platform technologies to evaluate undesirable metabolic perturbations in GM food crops.

A Post-genome Challenge: In our view the identification of unintended effects encompasses the ability to interpret and use innovations in molecular genetics research such as genomics, proteomics and metabolomics on a crop-by-crop basis. This will establish a whole data package directed towards a holistic view on possible unintended side effects due to genetic modification.

In essence, the focus is shifting towards molecular characterization to understand functional activity.

Vital to this approach will be informative profiles of molecules at different integration levels, *e.g.* mRNAs, proteins including post-translational modifications and molecules at the level of primary and secondary plant products and metabolites that might be useful benchmarks for the detection of unintended effects. The relative importance of these various 'profiling' techniques in establishing data sets for assessing unintended effects is not indicated by their order and will vary from species to species.

Alterations in, for example, expression levels, post-transitional modification, interactions or chemical composition do not per se imply that the product is less safe or even unsafe.

Such a platform of technologies may determine whether unintended effects exist between the GM plant and its control(s) or whether there is substantial equivalence apart from certain well-defined expected differences. Measures of the relative activities of various molecular constituents associated with metabolic capabilities in non- and modified plant cells under different conditions will provide new insights into how metabolism and genetic modification are orchestrated. But it should be emphasized that the fields of plant genomics and proteomics are still in their infancy.

CDNA MICROARRAYS

Over the past few years, the DNA microarray technology has emerged as a powerful high-throughput method for the analysis of gene expression. At present, microarrays for gene expression studies generally consist of cDNAs, which are physically deposited onto small glass surfaces. The major advantage of DNA microarray technology over conventional gene-profiling techniques is that it allows small-scale analysis of expression of a large number of genes in a sensitive and quantitative manner. Furthermore, it allows comparison of gene expression profiles under a large number of different conditions.

The cDNA microarray technology has already been successfully introduced in disciplines ranging from cell and developmental biology to drug development and pharmacogenomics. In the following a number of technical aspects of the technology as well as its potential value for the safety assessment of genetic modification of food plants will be discussed.

The cDNA microarrays are produced by deposition onto a solid surface of purified PCR products corresponding to specific genes. Typically a few nanolitres of DNA solution (100-500 μg/ml) are spotted using a micro-dispensing robot. Essential information on robot construction and protocols can be found on the Internet. In general, with respect to the spotting devices, a higher density of spots is accomplished when using a passive dispenser (> 2500 DNA spots/ cm2). Thereto, glass slides covered with a positively charged layer (*e.g.* poly-l-lysine) or carrying reactive groups are most commonly used. To analyse the expression of genes in a plant cell or tissue sample, polyA+ RNA (typically

0.5-2.5 μg is required for each reaction) is purified and labelled by reverse transcription using an oligo-dT primer incorporating a fluorescently labelled nucleotide. The labelled cDNA pool is then hybridized to the microarray. The required hybridization conditions (*i.e.* sample concentration, stringency of hybridization) and parameters, such as detection level and the correlation between transcript concentration and hybridization have been determined for each experimental setting.

The intensity of a hybridization signal, which is a measure of the relative expression level of each gene, can be read by charge-coupled device (CCD) cameras or confocal/non-confocal laser scanners. RNAs extracted from two different samples can also be simultaneously analysed with a microarray by labelling each with a different fluorescent label using either Cy-3 or Cy-5. At RIKILT-DLO we are currently testing whether the analysis of differential gene expression using DNA microarrays is an informative strategy for the safety evaluation of GM plants.

To address this question, DNA libraries are constructed and enriched by a subtractive cloning approach for cDNAs preferentially expressed in either green or red tomato fruit. These subtracted cDNAs, as well as control cDNAs representing known tomato genes (*i.e.* sequence information derived from public sequence databases), have been spotted on arrays and are used for comparison of gene expression profiling of control and genetically modified tomatoes. In the near future we will construct more defined arrays containing genes of toxicologically relevant pathways, such as those involved in the synthesis of natural plant toxins (*e.g.* the glycoalkaloids a-tomatine and a-solanine).

MRNA FINGERPRINTING

A very promising strategy to evaluate the equivalence of genetically modified crops is gene expression profiling. Conventional methods for the analysis of differential gene expression include Northern blotting, S1 nuclease protection, comparative expressed sequence tag (EST) sequencing, differential display and serial analysis of gene expression. More recently, Kok *et al.* have designed a method of detecting altered gene expression by means of mRNA fingerprinting or reverse transcription polymerase chain reaction (RT-PCR).

Based on the original concept of Liang and Pardee specific subsets of the mRNA population of a tomato plant were amplified. This enables the expression of different genes in the GM crop and genes in the parental line to be compared and any empirical information about significant alterations to be established. The method is still under development and requires further validation, but has been shown to be reproducible and able to detect important differences in gene expression.

Some of the most significant remaining problems are that the method largely depends on individual skills based on experience and is time-consuming, even in its most simplified form.

PROTEOME PROFILING

The introduction of Bt genes of bacterial origin may theoretically influence the nature of post-translation modification (*e.g.* glycosylation) in the transgenic tomato fruit. This is because as well as the impact of the protein moiety, for example, the type of glycosylation pattern may alter due to inactivation of glycosidases and glycosyltransferases in the host cell.

In addition, glycosylation of the inserted Bt gene product in the plant might result in altered protein activity and stability. The importance of structurally elucidating these oligosaccharides, especially the N-glycans, can be attributed to their widespread functions in the cell. These functions include correct folding, biological activity and stability of proteins as well as involvement in plant development.

Moreover, N-glycans represent antigenic epitopes by themselves. Immunological as well as biochemical studies revealed the allergenic potency for the á1,3-fucosylation of the proximal N-acetylglucosamine residue. Additionally, also the â1,2-xylosylation of the b core mannose is suspected to be an allergenic epitope. Recently, there is increasing evidence that these N-linked oligosaccharides may have a substantial impact on the immunogenicity of glycoproteins, as N-glycans are suspected to play a certain role in the adverse food reactions of hypersensitized patients. At the Institute for Agrobiotechnology, Austria, this issue was tackled by developing a generic technology for isolating, purifying and characterizing N-linked oligosaccharides in plant tissues.

Our low-temperature acetone powder method appeared to be most suitable for the isolation of (glyco-)proteins from plant tissue (*e.g.* tomato fruit). The resulting (glyco-)protein fraction was proteolytically digested and corresponding glycopeptides purified by cation-exchange chromatography. The oligosaccharides (N-glycans including á1,3-fucosylated oligosaccharides) were released from the peptide backbones by N-glycosidase A (PNGase A) and fluorescently labelled with 2-aminopyridine.

Subsequently, pyridylaminated glycans were separated and analysed structurally by repeated two-dimensional high-performance liquid chromatography (2D HPLC) using a size-fractionation and a reverse-phase column in combination with an exoglycosidase treatment.

MALDI-TOF (matrix-assisted laser desorption ionization time-of-light) mass spectrometry was applied for further confirmation of the observed structures.

Sixteen different N-glycosidic structures could be detected in tomato fruit. The two most abundant glycans in all the tomato samples showed identical properties to those of the major N-linked oligosaccharides of horseradish peroxidase (MMXF: Man(3)Xyl-FucGlcNAc(2)) and pineapple stem bromelain (MOXF: Man(2)Xyl-FucGlcNAc(2)), respectively, and accounted for about 65-78 per cent of the total glycan content.

Oligomannosidic glycans occurred in only small quantities (3-9 per cent). The majority of the N-glycans found was b 1,2-xylosylated and carried an á1,3-fucose residue linked to terminal N-acetylglucosamine. A structural element was identified and shown to be an IgE-reactive determinant, which contributes to cross-reactivities among non-related glycoproteins, and is present in a wide variety of plants extracts. It is proposed that this structural element might also be of importance in some cases of food and pollen allergy. The carbohydrate profiling of, for example, populations of green and red-ripe fruit of non-modified and modified Bt tomato plants (expressing the cryIA5 gene) revealed no differences in the nature of N-glycans, nor were there significant differences in the relative amounts of glycan. Interestingly, the relative amounts found were quite similar across seasons and comparable to the commercial variety Notoro used as an extended control. As far as the N-linked oligosaccharides were involved it was not possible to detect significant alterations in the á1,3-fucose content of Bt tomatoes due to genetic modification.

PROFILING OF THE METABOLOME

Among other factors, identity of a complex GM crop plant with its traditionally bred parent entails metabolism and the level of (un-)desirable substances contained therein. Therefore, we would like to build up an overall picture of components associated with different metabolic capabilities in the (non-)modified plant tissue.

A multi-compositional analysis of the so-called metabolome may overcome the test limitations of, for example, single critical nutrient or key toxicant analyses and problems related to their limits of natural diversity. An attractive alternative is the utility of high-resolution proton (1H) nuclear magnetic resonance (NMR) spectroscopy (1H-NMR) in combination with separation techniques (liquid chromatography (LC)). Recently, we have shown that the approach of chemical fingerprinting using off-line LC-NMR can obtain information on possible changes in complex plant matrices due to environmental effects as long as GLP-like conditions for sample handling, data acquisition and automation of data handling are maintained.

In principle, it will be possible to screen all the different low molecular weight components (MW <10 kDa) present in the metabolome of the plant tissue.

PRINCIPLES FOR ESTABLISHING A CHEMICAL FINGERPRINT

The chemical hashed fingerprint of a molecule is bit string (a sequence of "0" and "1" digits) that contains information on the structure. Chemical hashed fingerprints are mostly used. The principles for establishing a chemical fingerprint were directed towards the detection of alterations in compositions in 1H-NMR spectra arriving from two populations, for example, non-modified counterpart(s) and an engineered one. In this way the unprocessed parts of the plant used for human consumption were fractionated and analysed, *i.e.* in tomato comparison on the fruit level.

Genetically modified tomato varieties, such as the antisense RNA exogalactanase fruit, have been studied using this innovative technology. The spread of measurements of the individual profiles was comparable for the GMO and its controls. By subtraction of the 1H-NMR spectra the differences of the means were calculated at 99 per cent confidence intervals between the GM crop average (n = 8/batch per line) and its isogenic control average (n = 8/ batch per line). The difference of the means was defined here as all differences of the means of the amplitudes of individual NMR signals obtained from spectra of various populations. The differences of the means were used to quantify how the transgenic crop plant differed from its parental line. The normalized 99 per cent confidence intervals of the normalized means of fractions A-E after eightfold independent replicates demonstrated that differences in amplitudes exceeding at least 20 per cent could normally be detected. This is an adequate sensitivity to allow statistical evaluation, and is in line with recommendations of the Nordic Council. They proposed that if the average value of a parameter differs by more than 20 per cent an explanation should be sought. With the purpose of establishing those differences that may arise from metabolic effects linked to genetic modification the default statistical analysis used a mixed effects randomized block model.

Random block effects such as variations in the stage of cultivation, location, logistics or climate had a significant impact on the overall chemical fingerprints of cultivars. For example, chemical fingerprints varied much more between field sites and seasons than between replicates in one plot. A tomato background mean (*i.e.* crop mean value) was established by constructing an extended range of non-modified tomato references for comparison with natural variability, and to assess the impact of external factors (*i.e.* variations in season, climate, etc.). The crop mean value was used to quantify how the differences of the means between the transgenic crop plant and its control differed from the natural diversity of commercial counterparts.

To differentiate between compositional changes either due to genetic modification, genetic variability or environmental variations the chemical fingerprints were studied in a hierarchical approach.

Hence, the fingerprints of the GMO are compared to those of:

- The isogenic control line grown side-by-side under identical conditions in one plot;
- The isogenic control line grown side-by-side under identical conditions at multiple sites;
- An extended range of commercial varieties of that crop;
- The influence of downstream processing.

It was recognized that 45 per cent of all compounds of the control line harvested in different years varied considerably in concentration. Most differences were found in fractions A, C and D. A similar degree of impact on composition was observed in the case of processing temperatures (*e.g.* 100°C versus 70°C) on tomato juice.

These findings were compatible with the protease activity surviving the 70°C treatment. But in fruit harvested from various individual plants, divided into two populations, approximately 95 per cent of all constituents were practically identical in concentration. Also the determination of the differences of the means between transformant tEG to its isogenic control bred side-by-side in the same plot showed that 249 out of 3000 amplitudes (8 per cent) varied in concentration. Assignments by NMR indicated that, for example, citric acid was present in fraction A of the control at 1.4 times the levels found in the transformant.

However, in fraction A of the transformant glutamic acid and/or glutamine showed an increase of 1.3 times the levels found in the isogenic control. On the other hand, one single aromatic compound, a-lycopene, appeared to be present in fraction E of transformant tEG at 2.5 times the concentrations found in the parental line. No alterations were observed in ß-carotene levels.

These effects may be the result of the antisense downregulation of the tEG1A gene in tomato. In our view the natural diversity must be taken into consideration, however, when interpreting the biological relevance of any significant difference found in the differences of the means between the GMO and its parental line. The parental line was extended by entering non-isogenic commercial references (*i.e.* crop mean value). Although limited in the range of lines, the comparison of the crop mean value to the differences of the means showed that the contribution and magnitude of differences (*e.g.* citric acid, glutamic acid/glutamine, a-lycopene) diluted out.

It was recognized that in the agricultural practice the conditions of culturing, processing and storage of appropriate references will be mostly unknown and they may have a non-isogenic genotype. Notwithstanding these complicating random block effects we recommend that, for example, the chemical fingerprint should not be limited to the transgene and its isogenic parent bred under identical circumstances only, but should also include a sampling through multiple sites and a comparison with an extended range of controls of that crop. Otherwise it would be a too limited approach towards the establishment of the claim of substantial equivalence and it would neglect the ranges of natural variations in that crop. In future, if many more lines have to be compared with each other this approach may benefit from multivariate statistical methods.

SIGNIFICANCE OF DETECTED COMPOSITIONAL ALTERATIONS

The significance of detected compositional alterations may need to be further explored by in vitro toxicity assays in order to screen for potential adverse effects and for mechanistic studies. It may even be that the latter approach leads to a redesign, refinement or even replacement of (sub-)chronic rodent feeding trials. In vitro systems derived from organs and tissues from animals and humans, and various types of cultured recombinant cell lines, are successfully used for screening for the toxic potential of single compounds. However, their ability to screen for the potential toxicity of whole foods or extracts thereof has been insufficiently explored up till now, and has been examined by us within the framework of a tiered safety evaluation of GM plants.

The validity of using in vitro test assays for comparative testing of whole food products, food ingredients and extracts thereof depends on the sensitivity and selectivity of the toxic response induced upon exposure. However, since these models are not yet validated, they should be considered as early warning or alert-systems to identify potential changes in toxicity of the engineered product only.

THE MUCOSA OF THE GASTROINTESTINAL TRACT

The mucosa of the gastrointestinal tract is clearly a primary, potential site of action for the effects of novel foods. Therefore effort has been concentrated on the development of a battery of intestinal cell systems (*e.g.* IEC-6, IEC-18, Caco-2 and INT407 cell lines) with different endpoints for initial cyto- and genotoxicity testing. The choice of a suitable method of sample preparation was considered of utmost importance in assaying vegetables. Freeze-dried tomato fruit was extracted (10 per cent suspensions, w/v) in water, 0.9 per cent saline or chloroform:methanol (2:1, v/v) immediately before incorporation into the culture medium.

It was encouraging that aqueous and chloroform/methanol extracts of, for example, red-ripe tomatoes up to a maximum of 10 per cent suspensions (w/v), had little or no general toxicity once the pH (> 6.0) and osmotic pressure had been taken into account. No toxicity to the intestinal cells Caco-2 and IEC-18 under short-term conditions was detected in red-ripe, non- and antisense RNA exogalactanase tomato using alpha-glutathione S-transferase (a-GST) lactate dehydroyrenase (LDH) leakage, reduction of 3-(4,5-dimethylthiazol-2-yl)-2,5 diphenyltetrazolium bromide (MTT conversion), neutral red (NR) uptake and total cellular protein as in vitro endpoints. On the other hand, green tomato fruit displayed a pronounced toxic and cell-damaging effect. No statistical significant differences were found between transformants and isogenic control lines in terms of their activity in the assays.

Treatment of tomato tissue with pepsin (ratio pepsin to tomato protein 1:250, w/w) did not result in an increase of or loss of inherent cytotoxicity. Supplementary investigations with naturally occurring plant components, for example, quercetin, b -carotene, a-lycopene, genistein showed that only steroid glycoalkaloids, such as a-tomatine, exhibited a rapid toxic effect, whereas a-tomatidine was ineffective. It was noted that the in vitro models did show general matrix-related toxic effects. For example, a-tomatine diluted in a matrix of extracted red-ripe tomato tissue (final concentration of 5-10 per cent w/v) showed an increase of dose-response-dependent cytotoxicity in human Caco-2 cells (*e.g.* IC50 7.9 ± 2.1 mg/L versus IC50 33.9 ± 5.3 mg/L of a-tomatine only). Using the COMET assay, suspensions of freeze-dried tomato fruit up to a maximum of 10 per cent (w/v) did not exhibit DNA-damaging effects in a range of gastrointestinal cells derived from rats (IEC-6 and IEC-18) and humans (Caco-2, INT407). Genetic modification did not result in increased genotoxicity, but fruit extracts were able to suppress the DNA-damaging effects of the known genotoxins H2 O2 and N-methyl-N2 -nitro-N-nitrosoguanidine (MNNG).

Although these endpoints are important toxicological measures they do not indicate the molecular mechanisms of toxic damage. So far, pronounced adverse effects could only be observed with glycoalkaloids. The screening for subtle differences in toxic responses of other plant constituents appeared to be limited. However, therefore, the use of critical endpoints for such studies other than cytotoxicity should be clearly defined.

ASSESSING TRANSCRIPTION FACTORS

Transcription factors (TFs) are the main regulators of gene transcription. Thus, knowing the genes that are targeted by a specific TF is of utmost importance for understanding developmental processes, cellular stress response, or disease etiology. Chromatin immunoprecipitation coupled with deep sequencing (ChIP-seq) allows for measuring the genome-wide binding of TFs. There is a demand for in vitro models that not only determine the (geno-)toxic potency of plant components or complex mixtures, but also provide information on the mechanism by which that compound exerts its toxic effect.

As an in vitro profiling system the eukaryotic stress gene assay (*i.e.* CAT-Tox(L)) was assessed. This utilizes human HepG2 cells stably transfected with chloramphenicol acetyltransferase (CAT) reporter constructs. The aim of the CAT-Tox(L) assay is to collect and quantify molecular responses to cellular stress and toxicity that have a transcriptional component in their regulation. The assay employs an ELISA method to measure the transcriptional activity of the promoter or response element constructs.

The advantage is that many of the stress responses occur before any measurable cytotoxicity, thus allowing monitoring of stress pathways at sublethal levels. Since 14 unique transcription responses can be measured simultaneously, the CAT-Tox(L) provides comprehensive cellular stress profiles. The assay generates specific stress gene induction profiles for a variety of stressful and toxic plant components such as genistein, diadzein and sinigrin.

Aqueous extracts of red-ripe, non-and antisense RNA exogalactanase tomato fruit dosed up to 0.05 g per assay did not cause any cytotoxicity to the HepG2 transfectants nor were there molecular responses to cellular stress or toxicity. Extracts of non-modified and modified green tomato fruit showed cytotoxicity and induced the construct containing the xenobiotic response element (XRE). This gene-induction profile was not comparable to that of, for example, the reference molecules dioxin, 3-methylcholanthrene, benzo[a]pyrene or b -napthoflavone.

These polycyclic aromatic hydrocarbons were characterized with a co-induction of cytochrome P450 1A1 and glutathione-S-transferase Ya subunit-CAT fusion constructs. There was substantial evidence that the systemic fungicides fenarimol and bupirimate did not induce any of the genes in this assay. There was no promotor in the assay that could be linked to universal cellular stress. A-tomatine caused severe cytotoxicity at high concentrations and yet did not induce any of the genes (*i.e.* causing membrane damage).

However, it was observed that a-tomatine diluted in a matrix of red-ripe tomato tissue (final extract concentration 5 per cent w/v) elicited an increased cytotoxicity concomitant with a dose-dependent response profile involving XHF (collagenase involved in inflammatory reactions sensitive to mitogens and cytokine IL-1). So far, the approach of the CAT-Tox(L) model shows potential but there is a need for validation based on molecular mechanisms of plant constituents. Although the induction of these stress genes represents cellular perturbations, the mechanistic validations need to be extended before any conclusions can be drawn on the value of these findings. Certainly, the development of a database to explore relationships between a wide variety of compounds stressful to plants and meaningful stress gene profiles may serve as molecular fingerprints unique for a particular component.

RESISTANCE TO PLANT VIRUS DISEASES

PROTEIN

One successful application of genetic engineering in crop improvement is in the production of virus-resistant plants. This has been achieved by genetic transformation of plants with novel resistance genes based on nucleotide sequences derived from the viruses themselves or from virus-associated nucleic acid as well as genes from other sources. The demonstration that the expression of tobacco mosaic virus (TMV) coat protein (CP) in transgenic plants protects the plants against virus infection has led to explorations of using other viral or virus-associated genes in producing genetically engineered resistance against in plant virus. Transgenic plants expressing Sat-RNAs have been obtained in a number of labouratories and have been tested in greenhouses and fields. Recent studies with defective interfering RNA or DNA protection and sense-antisense-ribozyme RNAmediated protection offer much promise for a broad range virus resistance in plants.

Our labouratory has successfully produced satellite RNA-mediated resistance to cucumber mosaic virus (CMV) in tobacco and tomato, CP-mediated resistance to rice stripe virus in rice and fusion protein (CP and nuclease) mediated high resistance to tobacco mosaic virus in tobacco. One way to confer more effective and durable field resistance to virus disease in transgenic crop is to use a combination of multiple resistance genes.

ENGINEERING RESISTANCE TO PLANT-VIRUS DISEASES

Since 1981, Sat-RNAs have been used as a biological control agent (BCA) of diseases caused by CMV on a large scale in China. The results show that the BCA is effective in controlling diseases caused by virulent CMV strains in many crops and, in addition, induces resistance to certain types of fungal diseases. The protective effects of CMV Sat-RNAs have been demonstrated in a number of greenhouse and field tests. These studies have provided the basis for engineering plant resistance to virus infection by introducing the Sat-RNA as a transgene.

In our labouratory, the cDNA of an attenuating Sat-RNA monomer from CMV strain 1 was synthesized and constructed into the plant expression vector pRok2. The Sat-RNA cDNA was introduced into different plant species including tobacco and tomato.7 When transgenic tobacco and tomato plants were tested in the greenhouse for resistance against challenge infection by CMV, they showed a significant decrease in disease index compared to non-transformed plants. In an experimental scale field test conducted in the spring of 1990, the transgenic tomato plants inoculated artificially with CMV gave about 50 per cent higher fruit yield than the non-transformed control plants and showed a decrease in disease index. However, there was no significant difference in leaf yield between the satellite transgenic and the control tobacco plants, although the disease index of transgenic tobacco was about 10 per cent lower than that of the control. Due to an insufficient level of Sat-RNA in the plants at an early stage of infection, the transgenic tobacco plants showed only a weak resistance to CMV, which directly affected the leaf yield, since it is the early leaves that are harvested in tobacco production. Although Sat-RNAs confer some extent of resistance to virus infection, field tests carried out in China from 1990-1992 show that the resistance of the transgenic plants is not strong enough to fully protect tobacco and tomato plants from the damage caused by natural virus infections.

SAT-RNA IN A MULTIPLE GENE STRATEGY OF ENGINEERING RESISTANCE

One way to obtain more effective and durable resistance to virus diseases in transgenic plants may be to use a combination of multiple resistance genes. In this strategy, different genes that produce blockage or interference at different stages of a virus infection are used to obtain engineered resistance.

Transgenic lines of tobacco expressing both CP and Sat-RNA of CMV were obtained through transformation using a chimeric vector containing two expression cassettes of the two genes under the control of the CaMV 35S promoter. When challenged with CMV, the virus concentration of challenging CMV in transformed plants that express only Sat-RNA was about 10-20 per cent of that in non-transformed plants, while in CP + Sat-RNA transformed plants it was only 4-5 per cent. A comparison of disease incidence and disease index of CP + Sat-RNA, CP alone or Sat-RNA alone transformed tobacco plants. The resistance of CP + Sat-RNA transformed plants was about twice as strong as that of Sat-RNA alone or CP alone transformed plants. Since the CP level expressed in CP + Sat-RNA transformed plants was similar to that in transformed plants that contained only the CP gene, the two-fold increase in resistance of CP + Sat-RNA transformed plants was not due to higher levels of CP expression. On the other hand, the level of Sat-RNA accumulation in the CP + Sat-RNA transformed plants was lower than that in the Sat-RNA alone transformed plants.

This may be due to the effect of the CP gene conferring resistance at an early stage in CMV infection, reducing the replication of CMV as well as Sat-RNA. Preliminary data from a field test of 1992 confirmed the enhanced resistance of the CP + Sat-RNA transformed plants to CMV. The CP + Sat-RNA transformed plants show stronger resistance to natural CMV infections than either the CP alone or Sat-RNA alone transformed plants.

TRANSGENIC TOBACCO PLANTS RESISTANT TO CUCUMBER MOSAIC VIRUS

Since *Nicotiana tabacuum* is an amphidipoid species, the tetraplold transgenic plants will segregate extensively in their progenies. It needs to be propagated several generations before obtaining a homozygous line. In order to speed up this process, a procedure for a fast production of homozygosity of transgenic plants was devised in our work. Leaf discs of diploid tobacco plant derived from anther cultures of a commercial cultivar NC-89 were transformed with a chimeric vector of CP and Sat-RNA genes of CMV.

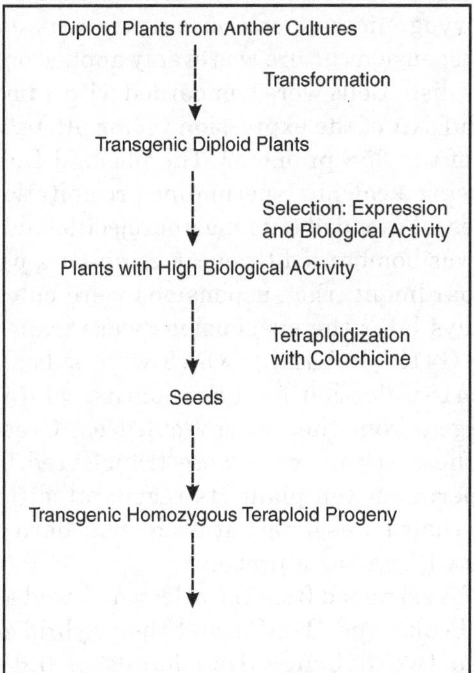

Fig. A Procedure for fast Homozygosity Generation in Transgenic Plants.

The kanamycin resistant diploid plants were selected by their expression of CP and Sat-RNA. Then the transgenic plants were tested for their resistance to CMV after inoculation with high concentration of CMV (50-100 ìg/mL). Only the plants with high resistance to CMV were tetraploidizated with colchicine. The transgenic homozygous tetraploid progeny can therefore be obtained after only one generation.

A certain amount of transgenic seeds will be produced within one growing season. After propagation of the transgenic seeds once in Hainan Province a large scale of field test can be achieved in the main tobacco producing area in the second year.

RESISTANCE TO RICE STRIPE VIRUS IN TRANSGENIC INDICA RICE PLANTS

Rice stripe disease occurs in rice growing areas of China, Japan, Korea and the Commonwealth of Independent State, and causes significant reduction in rice yield. The coat protein (CP) gene of a Chinese isolate of rice stripe virus (RSV-C) was synthesized, cloned and sequenced, which is very similar to the Japanese strain. Cell suspension cultures were initiated from embrogenic calli of rice Annong S-1 (indica rice) by inoculating yellowish, compact and embryogenic calli derived from seeds into a suspension culture medium containing proline and maltose.

After being cultured at 26°C in the dark for about half a year, finely-dispersed and embryogenic suspension culture was established. Before bombardment, the suspension culture was evenly applied onto three-layerfilter-paper discs in a petri dish. Cells were bombarded with 1 ìm diameter tungsten particles coated with DNA of the expression vector pROK2 containing the CP gene under control of the 35S promoter. The plasmid DNA was absorbed to tungsten particles using a calcium-spermidine precipitation procedure. 2.5 ìL of the coated particles was loaded onto macroprojectile and each dish with the suspension culture was bombarded three times under a partial vacuum.

Following bombardment, the suspensions were cultured in a modified N6 medium. Two days later the suspensions were transferred to the same medium containing G418 (40 ìg/mL), which were subcultured every week. Being subject to G418 selection for two months, white and fast-growing colonies were emerged from the browny cultures. Green plants could be regenerated when the resistant calli were transferred to a differentiation medium. In one experiment ten plantlets regenerated from G418 resistant calli were tested for their transgenic nature by Southern blot analysis using a32P-dCTP-labelled CP gene as a probe.

The genomic DNA extracted from the selected as well as the control plants were digested with EcoR1 and BamH1 and then hybridized with the probe. Two plants showed two hybridization bands of 0.6 kbp and 0.7 kbp corresponding to the CP gene. Western blot and ELISA further analyses demonstrated that CP (32 kDa) was expressed in the transgenic rice plants.

Sixteen transgenic thus obtained and 100 non-transformed rice plants were inoculated with RSV-carrying leaf hoppers, 24 days post inoculation feeding, most of the untransformed plants developed chlorotic stripes and chlorosis in the young leaves, the disease index (DI) being 53 per cent.

After 40 days post inoculation feeding, most of the control plants showed severe symptoms, being stunted and the leaves developing chlorotic and brown necrotic streaks, with the DI being 88 per cent. In contrast, the transgenic rice plants subject to the inoculation feeding developed no or mild symptoms. After 24 and 40 days post inoculation feeding, the disease index of the transgenic plants was 15 per cent and 25 per cent respectively.

INCREASED RESISTANCE TO TOBACCO MOSAIC VIRUS

Coat protein-mediated resistance (CPMR) has been proved to be successful against several families of plant viruses. Recently CPMR was obtained by expressing a truncated CP of tobacco virus. Our work has indicated that expression of a fusion protein of TMV-CP and Staphylococcal nuclease (SN) in transgenic tobacco plants significantly enhances the CPMR to TMV. TMV-CP or SN gene alone and CP'!SN or SN'!CP fusion protein genes were obtained by PCR, inserted downstream of 35S promoter of a binary vector pROK2 and transformed into tobacco leaf discs.

No SN gene alone transformed tissues could be regenerated, but the tissues containing CP'!SN or SN'!CP fusion genes grew normally and could be regenerated into plants. The transformed nature of the regenerated tobacco plants were confirmed by Southern blot hybridization and PCR analysis. It was demonstrated that the fusion proteins of CP'!SN and SN'!CP were expressed in the transgenic plants by Western blot analysis. CP'!SN- or SN'!CP-transformed plants exhibited higher resistance to TMV than CP alone transformed ones. 50 per cent of CP'!SN- and SN'!CP-transformed tobacco plants were symptomless for three months after inoculation with 1 ìg/mL TMV.

Nuclease activity test of the fusion protein expressed in transgenic plants was carried out *in vitro*. Fusion proteins of SN and CP were extracted from the transgenic plants and were obtained by immunoprecipitation with antiserum of TMV. The precipitated proteins were infiltrated with 2 per cent SDS and then with water containing excess amount of TMV-CP, and reacted at 30°C with the incubation mixture of 25 mM Tris-HCl (pH 8.8), 10 mM $CaCl_2$ and 40 ng 32P-5'-end-labelled oligdeoxynucleotide. The result showed that no nuclease activity was observed. No obvious difference in structure and infectivity of virion isolated from SN-CP transgenic and non-transgenic plants has been found.

The mechanism of resistance to virus by expression of a fusion CP was unknown. We speculate that the fusion CP is somehow dysfunctional and more effective to disrupt the normal virus-host relationship than CP does alone. The disruption may be at the level of virus movement throughout the plant or the fusion CP is encapsidated into virion to generate defective virus particle.

RISKS ASSESSMENT OF RELEASING GENETICALLY ENGINEERED PLANTS

Some transgenic organisms, such as micro-organisms and animals will be grown under controlled conditions and only released to the environment accidentally. However transgenic plants will be grown on a large scale in the open environment. There is concern that transgenic plants may produce risks.

Spread of Transgene by Pollen

Because genetic engineering will often result in the introduction of genes not previously present in a species, there is concern that such genes may spread to other plants and to weeds by crosspollination. It is important to have information on pollen transfer distances and the range of species that can be pollinated.

To obviate this problem trial release of transgenic plants have often required the removal of flowers to prevent pollen spread as for the transgenic tobacco production. However, if such plants are to be used for the seeds or fruits, unlimited spread of pollen will occur.

Releasing of Antibiotic Resistance Genes

Antibiotic resistance genes are often used as a selective marker in transformation procedure. In the case of antibiotic resistance, it will be a problem only if it is inherited by pathogenic micro-organisms which are being controlled by the antibiotic in question. The risk of a pathogen inheriting and being able to express an antibiotic resistance gene derived from a plant is at least a million-fold lower than such resistance being obtained from other micro-organisms.

Pollution of Honey

A further problem that might be encountered is the presence of cloned gene products in pollen. This could harm the bee larvae which are fed pollen, pollutes honey with minute amounts of the gene products, and makes the pollen more allergenic.

Most gene products will not be found in or on pollen in other than minute amounts and thus contamination should not be a problem.

Transgenic Plants Becoming Weedy

Potentially the greatest risk is that certain types of transgenic plants might become weedy, for example, the acquisition of herbicide resistance genes which will make them less easy to control in subsequent generations. On the one hand herbicide resistance genes are used as selective markers in many labouratories; on the other hand large companies are attempting to encourage growers to use more herbicides, and are using genetic engineering for purely commercial reasons.

Safety of Application Sat-RNA for Virus Diseases Control

Safety concerns about the use of Sat-RNAs remain. Even though virus replication is inhibited in transgenic plant, a small amount of Sat-RNA can still be encapsidated into helper virus particles and transmitted to other plant species, which it may cause more severe diseases. Moreover, because the exacerbating and attenuating Sat-RNAs can differ in only a few nucleotides, there is always a possibility that a symptom attenuating Sat-RNA can mutate into an exacerbating one.

During a decade of using Sat-RNAs as biological control agents in China, a number of facts concerning the safety of using Sat-RNA in agriculture have been realised. In spite of the large scale application of the BCAs on different crops, no necrogenic or other harmful Sat-RNA variant was found in a systematic field survey in which random plant tissue samples were analysed by temperature-gradient gel electrophoresis (TGGE) to detect Sat-RNA sequence variants.

Secondly, the efficiency of aphid transmission of CMV in the presence of the Sat-RNA is reduced to about 10 per cent of that in the absence of the Sat-RNA.

The reduced transmission efficiency, due to the low concentration of CMV when the Sat-RNA is present, lowers the chance that the Sat-RNA will be released in the field. Finally, it is possible to maintain the benign property of the Sat-RNA by periodic and detailed quality control of the BCA stocks.

It is our hope that the use of Sat-RNA as a viral control strategy will become increasingly safe as our knowledge of Sat-RNA structurefunction relationship increases. Indeed, it has recently been found that the Sat-RNA of CMV can lose its ability of being encapsidated into CMV CP by changing a few nucleotides; this in turn excludes the possibility of Sat-RNA transmission (D. Baulcombe, personal communication).

Safety of viral coat Protein-mediated Resistance

Transencapsidation of viral nucleic acid by transgenically expressed CP may induce heteroencapsidated virion, which might be transmitted by new vectors and to new host plants. The CP of plum pox potyvirus has been shown to confer aphid-transmissibility on a non-transmissible isolate of zucchini yellow mosaic virus. However, recent studies showed that intentionally truncated, antisense or non-expression (-AUG) CP genes could provide measurable protection or even complete immunity against the appropriate parent virus; this in turn excludes the possibility of heteroencapsidation of viral nucleic acid by transgenically expressed CP.

In conclusion, safety, from a public view, was ensured by biologically containing the transgenic plants so that they would not induce risk. The emphasis on biological containment has meant that for release to be permissible it is necessary to demonstrate beyond reasonable doubt that the transgenic plants are safe. But nothing is completely safe, and safety *per se* is not always desired; for example, plant species without any toxic metabolites would be exquisitely sensitive to pests.

GENE EXPRESSION PROFILING

Gene expression profiling simultaneously compares the expression levels of many genes between two or more sample types. This analysis can help scientists identify the molecular basis for phenotypic differences and select gene expression targets for in-depth study using other technologies. Gene expression profiling provides valuable insight into the role of differential gene expression in normal biological and disease processes. A second potential application of DNA microarray technology in safety evaluations concerns screening for biological effects of (fractions of) plant compounds that have been found by other profiling techniques to be present or modified in the genetically modified crop plant when compared to the parental line and other references. At RIKILT-DLO we are testing the feasibility of this strategy by analysing defined food compounds for possible effects on human intestinal gene expression.

To this end, human intestinal cell lines are exposed for different time periods to various amounts of crop plant components and the mRNA isolated. The labelled mRNA is hybridized to DNA microarrays containing unknown intestine-specific genes, as well as already identified control genes known to be intestine specific or to be functional indicators (for toxicological processes, apoptosis, etc.).

As a source for unknown, intestine-specific genes we use cDNA libraries from human intestine biopsies and human intestinal cell lines which have been enriched by a subtractive cloning approach. Having established the usefulness of this approach for the identification of biological functions of defined plant compounds, these cDNA microarrays will be exploited to screen, for example, GM crop plants for possible effects on human health.

The concept of substantial equivalence is broadly accepted as a basis for the risk assessment of GM crops and derived novel foods. Data requirements for establishing substantial equivalence should be based on identifying unintended metabolic perturbations in the engineered crops that have been caused by genetic modification. A generic approach cannot be envisaged, however. In all cases the collation of data should be directed by the type of genetic modification and related consequences. It is recommended that unintended effects should be identified on a case-by-case basis, since such effects are dependent on the genotypic and phenotypic parameters of the new gene products involved. The application of innovations in molecular genetics research will help to define the conditions under which the new food products can be marketed. The present approach has generated a sound scientific basis for the evaluation of the potential of unintended effects in GM crops and derived food products.

The results may assist regulators and legislators to test and, if necessary, to improve the applicable regulations. Moreover, the technologies described and the results obtained may serve as a general framework for the risk assessment of GM crop plants, and may contribute to a better understanding by the public of recombinant DNA techniques in plant breeding and their implications. It is concluded that a parallel approach focusing on a comparative analysis using informative profiles of molecules at various integration levels (*e.g.* mRNA, protein, metabolites) including a toxicological profiling of relevant plant matrices, offers good prospects for the identification of hazards related to unintended effects.

The choice of comparators and of external parameters to support a claim that a GM crop plant presents no unintended effects compared with its traditional counterpart should be based on sound scientific judgement. It is recommended that data are generated on the unprocessed part of the plant, such as the tomato fruit or potato tuber, used for human and/or animal consumption. Furthermore, data banks should be set up containing information on natural variations of essential plant or other product constituents (*i.e.* crop mean values).

The chemical fingerprinting technique, using a combination of off-line liquid chromatography and proton-NMR imaging, appears to be a powerful screening method for the detection of secondary effects in the metabolome that may be applied on a routine basis. The carbohydrate profiling technique is useful for the detection of unexpected changes at N-glycan levels; in particular it is applicable to the detection of post-translational modifications in newly expressed proteins as well as in the whole GM crop plant.

DNA microarray technology is rapidly evolving and much effort is being put into improving spot density, reduced production time, and increased reproducibility and sensitivity. Improving the latter is critical because it makes it possible to use smaller amounts of starting material. It will depend on technical aspects such as the quality of the scanners and spotting machines, the development of fluorescent dyes with improved characteristics (*e.g.* narrow excitation and emission peaks; high level of photon emission; resistance to photo-bleaching), supports with reduced background and more target sequence binding capacity.

DNA microarrays generate a huge amount of complex hybridization data and major challenges are to develop more advanced computer software to help to find statistically significant correlations within, and between different experiments, and to link the data to sequence information and submitted expression profiles available in public databases.

Application of these technologies in combination with other measurements of, for instance, the performance and quality may limit or even replace animal feeding studies aimed at the detection of unintended effects. Together, these techniques, known as functional genomics, will enable us to answer questions about what happens to expression and protein composition when a plant cell changes metabolic state, for example, due to genetic modification or environmental factors. It is foreseen that this holistic view on the cellular machinery will be achieved soon, as the coincidence of genome sequencing with improvements in the analysis of expressed proteins is of growing importance and continues to be successful. Eventually, microarrays for genomic studies will probably also be used to evaluate the numerous constituents of GM crop plants. Some constituents might not cause obvious changes in cellular behaviour or morphologic appearance but could cause subtle metabolic alterations that would show up when the mRNA content was interrogated by an array. It is likely that microarrays might be used to evaluate the life cycle of a plant much more precisely and to understand the complex metabolic control systems of plants.

The ultimate test for GM crop plants will be acceptance by the public in large. It is therefore extremely important that industry, the scientific community and the regulatory authorities are able to respond to queries from the public by providing transparent information on the criteria for safety assessment. Improved safety assessment at the molecular level should refine and complement strategies by anticipating the potential risks and sources of unintended effects in GM plants.

6

Gene Biotechnology in Medicine

INTRODUCTION

Biotechnology has wide range of applications in animal and human health care both directly as well as indirectly. It's direct use involves controlling the spread of a variety of diseases using gene manipulation at different levels. The indirect use is in manufacturing of nutritional food through single cell proteins and in detecting food contaminants through molecular probes.

Following approaches are used to control the diseases and thereby taking care of the human and animal health.

- Providing immunity against the disease through the development of vaccine.
- Early diagnosis of a disease.
- Successful treatment of the disease.

Development of vaccines for immunity:

Biotechnology is used in three different ways in the development of vaccine:

- Separation of a pure antigen using a specific monoclonal antibody.
- Synthesis of an antigen with the help of a cloned gene.
- Synthesis of peptides to be used as vaccines.

USE OF MONOCLONAL ANTIBODIES FOR IMMUNOPURIFICATION OF ANTIGENS

The method of immunopurification using monoclonal antibodies is used, to separate specific antigen from a mixture of very similar antigens. Once purified, the antigen is used for developing vaccine against a pathogen. Individual interferons (which have the property of inhibiting viral infection and cell proliferation) have been purified using this technique. These interferons were later used for clinical trials and then commercially used.

Use of Cloned Genes for the Synthesis of Antigens

Hundreds of genes in eukaryotes have been cloned from genomic DNA or from cDNA. These clones genes included a number of genes for specific antigens and some have been used for the synthesis of antigens leading to the preparation of vaccines. A very good example of this is cloning of Hepatitis B virus (HBV) genome. The HBV genome was cloned in the plasmid pBR322 followed by it's propagation in *E.coli*. The antigens produced from this clone reacted with hepatitis B core antibody (HBAb) which has been used to produce hepatitis B vaccine.

Efforts are on to use this method to produce an Anti- malarial vaccine. Malarial parasite *Plasmodium falciparum* is the malarial parasite which has become a major threat to human health which spreads by the mosquito bite. In human body, the malarial parasite passes through several antigenically distinct phases *viz.* sporozoite, merozoite, gametocytes etc. The vaccines can be developed to control any of these phases hence we can have antisporozoite vaccine, antimerozoite vaccine etc.

Out of all these, considerable progress has been made in the making of antisporozoite vaccine due to cloning of gene meant for circumsporozoite (CS) protein. This protein was obtained directly from DNA of erythrocytic form of parasite, rather than as cDNA from mRNA.This cloned gene may, in course of time, lead to the synthesis of vaccine by synthesizing CS protein by clone gene.

Synthetic Peptides as Vaccines

Vaccines can also be prepared through short synthetic peptide chains. There are several ways by which these can be used as vaccines. As it is the three dimensional structure (not the amino acid sequence) of the protein which is responsible for the immunogenic response, it is essential to find out the protein region involved in immunogenic response. *E.g.* in Foot and Mouth Disease virus (FMDV), the amino acid 114-160 of virus polypeptide can produce antibodies which can neutralize FMDV and provide protection. The region of 201-213 amino acids of the same protein also could neutralize FMDV hence it has been shown that small synthetic peptides representing these regions of proteins can show immunogenic response and can be used for the development of vaccine.

The immunogenic region of protein can also be located by gene coding for the protein. *E.g.* in Feline leukaemia virus, the clone gene of an immunogenic protein was cut into fragments by DNAase I and then cloned in lambda phage. Phage colonies (plaques) with different cloned fragments are screened with a specific monoclonal antibody that neutralizes the pathogen. The fragments which react with antibody must be synthesizing the immunogenic peptide fragments which can be sequenced. Using this method it was possible to identify a 14 amino acid immunogen of the envelope protein of Feline leukaemia virus (FLV). The corresponding synthetic peptide was also found to compete with the virus for antibody.

The immunogenic region of a protein in a pathogen can also be identified by eluting it from purified major histocompatibility complex (MHC) molecules. Different MHC allelic variants bind with different proteins and purified using specific T cells. Peptides can be eluted from these purified MHC molecules and later on sequenced. The sequenced peptides are used to make synthetic peptides which are used as vaccines.

DIAGNOSIS OF DISEASES

DIAGNOSIS OF DISEASES CAUSED BY PROTOZOA AND HELMINTHES

The monoclonal antibodies and DNA probes are being used as very sensitive tools in biotechnology to diagnose the diseases caused by protozoa and helminthes. Monoclonal antibodies can be used through serological tests which takes only minutes as compared to conventional methods which require some weeks as the bacteria and viruses have to be cultured *e.g.* in Herpes virus.

The DNA probes are more sensitive than monoclonal antibodies and the process takes hours instead of weeks. Readymade DNA probes for Herpes virus and other human, animal and plant viruses are being prepared. Probes are now available for a number of human parasites from the group protozoa and helminthes.

Diagnosis of Sexually Transmitted Diseases (STD) using Monoclonal Antibodies

The occurrence of sexually transmitted disease has gone up in recent years both in developed as well as developing countries. Some of the pathogens causing these diseases are- *Neisseria gonorrhoeae, Chlamydia trachomatis, Herpes simplex virus* etc. The conventional methods for the diagnosis of these diseases are microscopic examination, culture method, Immunological method, and measurement of antibody.

The use of monoclonal antibody has been very effective in the diagnosis of these diseases. The cell hybrids are prepared using hybridoma technology and screening of cell lines is done through replica plating technique. Hybrid cells are placed in a 96-well microtest plate. Small sample of culture fluid from each well is placed on replica plates followed by impregnation of each well with a specific antigen. The immune reaction is detected through radio-immunoassay in which 125I-labelled protein A (which binds to Fc protein of human IgG in the immune complex) is added to each well and then examined by autoradiography. The positive reactions can be traced to original wells and the cell line then multiplied for production of specific monoclonal antibodies. Using this technique, monoclonal antibodies could be prepared that would distinguish between the different pathogens causing STDs.

Antenatal Diagnosis of Congenital Diseases

If there is a chance of giving birth to a baby with congenital defects then it may be desirable to diagnose the condition of the foetus during gestation.

This can be done by extracting some cells from the foetus by drawing a few milliliters of amniotic fluid with the help of hypodermic needle. The free cells of the foetal origin from the amniotic fluid can be cultured and tested for karyotype analysis, enzyme production, and restriction fragment analysis of it's DNA, to detect any abnormality.

It is possible to identify a disease with in two months of pregnancy as compared to 18 weeks period required earlier. The number of disease specific DNA probes is also increasing at a fast rate which will further help in antenatal diagnosis.

TREATMENT OF DISEASES

MANUFACTURING OF PHARMACEUTICAL DRUGS THROUGH BIOTECHNOLOGY

Various drugs are being manufactured using biotechnology to treat different diseases. Using gene therapy, two drugs –Insulin and Interferons have already been produced. Insulin is used to treat diabetes and Interferon is used for the treatment against some tumour viruses. It is possible to manufacture these drugs in large quantities by cloning the corresponding genes from human or animals through plasmid vectors in bacteria. This procedure brings down the cost of manufacturing the drugs. Lots of research is also going on a number of proteins such as urokinase, factor VII:C, human growth hormone (HGH) etc. The Human Growth hormone for treating dwarfism due to hypopitiuitary activity was synthesized using recombinant DNA technique and was commercially introduced under the market name of prototropin in USA and somatonorm in Britain. Both Insulin and growth hormone were manufactured under license from Genetech Inc. based in USA.

Gene Therapy

If a child or an embryo is diagnosed to carry a defective gene leading to disability, following methods are used to correct this defect:

- Replacement of defective gene with a normal gene.
- Correction of the defective gene through gene targeting.
- Gene augmentation through increasing the number of copies of the gene or through a higher level of expression of the introduced gene.

All these methods used to correct defective gene is called gene therapy. Out of all these methods, targeted gene modification for gene correction or gene augmentation by introducing normal foreign gene sequences are being widely used in gene therapy.

Several reports are already published where targeted gene modification has been demonstrated in mammalian systems. The genes are introduced by gene transfer methods such as electroporation, transfection etc., followed by site specific mutagenesis *e.g.* HGPRT locus and int-2 loci in mouse embryonal stem cell. In past few years, approved gene therapy experiments have been conducted of which the transfer of Neo R/TIL gene marked immune cells into patients with advanced cancer was successful. ADA gene therapy used for treating the deficiency of adenosine deaminase deficiency (ADA) involved the transfection with lymphocytes bearing the ADA gene carried by retroviral vector. In Cancer gene therapy, Tumour necrosis factor gene (TNF) or IL-2 gene is inserted in TIL tumour cells isolated from the patients and grown in culture. After this these gene corrected cells are injected into the body of the patient.

In gene augmentation method, a number of copies of the desired gene are introduced and are made to express at high level using expression and transfer vectors. After the gene correction at cellular level, the implantation of the modified cells into a suitable region in an organ of the patient or in the embryo is carried out.

Therefore the gene therapy can be used at two different levels:

- Embryo therapy
- Patient therapy.

Embryo Therapy

In this, the genetic constitution of embryo at post-zygotic level is altered which leads to the alteration in inheritance as well.

Embryo therapy involves the following steps which have been tried in case of mouse or rabbit only:

- *In vitro* fertilization of the egg.
- Insertion of normal gene into embryo at post-zygotic level either using viruses or directly by microinjection.
- Integration of inserted gene in host dna, where it may or may not function.

The therapeutic newly inserted gene may or may not function under normal control in the animal, in time, space and quantity.

Patient Therapy

In patient therapy the cells with healthy genes are introduced in the affected tissue but the inheritance trait of the patient is not affected or altered.

This therapy involves the following steps:

- Identification of defective gene
- Isolation or synthesis of normal healthy gene.
- Isolation of the cells of the tissue where the normal healthy gene will need to function.
- The placement of normal gene into a cell where it can function.

Drawbacks of this Method

First there is the possibility that the introduced gene may not function, and the second is when corrected cells are reintroduced, these maybe outnumbered by the diseased resident cells. The other problem is that there are only few diseases affecting only a single tissue. During the last few years, the genes are being routinely isolated. The isolated gene may either be directly injected into the cell or be carried by a virus, to which it is linked by recombinant DNA technique. After entering the cell, the gene either remains free in cytoplasm like extrachromosomal DNA or become a part of nuclear DNA. However under both circumstances only few copies of RNA are synthesized as compared to thousands of copies of RNA per cell by normal cells.

Gene therapy holds a promising future and hope for patients suffering from diseases which can be treated using this technique.

CULTURED STEM CELLS AND BONE MARROW TRANSPLANTATION

Stem cells are undifferentiated cells which have the potential of differentiating into several kinds of cells. These cells are used for transplantation in patients, whose immune and blood forming systems have been devastated by leukemia, cancer, chemotherapy or other unknown causes. In the foetus, the blood cells are created in the liver, but just after birth they are produced exclusively in bone marrow. If marrow is transplanted to the same individual from where it was taken for culture, it is described as autotransplantation and if donated to someone else, it is called allogenic transplantation.

Monoclonal antibodies have been used to stem cells from the mass of cells in the bone marrow. These monoclonal antibodies react to a cell marker or antigen called CD34, which is specific for primitive bone marrow stem cells.

Several hormones are also available which stimulate replication of precursor blood cells *e.g.* granulocyte colony stimulating factor (G-CSF), Interleukin-6, Interleukin-11 and stem cell factor.

Monoclonal antibodies linked with toxins can be used for killing the malignant cells. Chemotherapy with extracted bone marrow can also be used for destroying malignant stem cells.

In future, efforts are being made to store the stem cells of a healthy person permanently and use them whenever required. This will be possible when the techniques to culture stem cells would be standardized specially using umbilical cord blood which may become an important source of stem cells.

Skin Related Ailments and use of Cultured Cells

Efforts are all ready on to use artificial skin derived from cultured cells to treat different skin related ailments especially severe burns. Skin is one organ that can be aritificially made from cell culture and used for grafting after skin damage caused by severe burns. Keratinocytes make 90 per cent of the epidermis of the skin can be cultured under *in vitro* conditions. Their proliferation is facilitated by the products of fibroblasts which are found in the dermis layer of the skin therefore Fibroblast cells called 3T3 cells were added in the culture of keratinocytes besides adding epidermal growth factors, cholera toxin and a mixture of growth factors. Only 1-10 per cent of epidermal cells grow as the others start differentiating. In order to stop stratification of cell layers, the cells are continuously separated and re-cultured until a sheet of pure epithelium is obtained.

The cells of this sheet of epithelium are linked by Desmosomes therefore it is detached from the vessel by using the enzyme dispase. After cleaning it of extraneous proteins, it is attached to a gauze and brought to the hospital for using it as graft.

The cultured keratinocytes have been used in:

- Removal of scars on skin using cultured cells.
- Regeneration of epithelium of mouth using cultured oral keratinocytes.
- Repair of congenital penile defects using cultured urethral keratinocytes.
- Treatment of several diseases related to middle ear.
- Treatment of chronic skin ulcers using cultured grafts.

Intensive research is going on to use cultured keratinocytes in various other areas such as using them as lining of vascular prostheses (artificial limbs), to repair urinary tracts, use of pancreatic islet cells to treat diabetes, use of cultured myloblasts to treat muscle diseases.

GENETIC COUNSELLING

Biotechnology has also contributed in the field of genetic counseling especially to the couples who may be at high risk of giving birth to a child with congenital defects or diseases. These parents can decide voluntarily not to have a child or may undergo selective abortions after investigations during antenatal care.

The hereditary or congenital defects are tested through the process of amniocentesis. Around the fifteen week of pregnancy, a sample of amniotic fluid containing fetal cells is removed and cultured in vitro. The cultured cells are then subjected to biochemical enzyme test and karyotype analysis. The results are particularly relevant to the situation where one or both parents have a family history of a particular hereditary disease and one has to evaluate the possible chances of the children having the same disease.

This type of test followed by genetic counseling is also done when a couple may have an abnormal child and would like to know the chances of having abnormal child on the next pregnancy. In some cases, if the defect is controlled by single recessive gene and both parents are normal, in three out of four cases the child would be normal. At least 35 diseases which can be identified by this technique are known. The number of disease specific DNA probes is also increasing at a fast rate, so that antenatal diagnosis by DNA analysis or 'enzyme linked immunosorbent assay'(ELISA) should be possible for all single gene defects *e.g.* the incidence of Thallassemia in Cypriot community in U.K. has decreased to 20 per year following the use of antenatal diagnosis.

Making a Choice of Baby's Sex

It is always interesting for a married couple to know the sex of the unborn baby. The chromosome techniques have made it possible to know the sex of the developing fetus by drawing amniotic fluid and preparing karyotypes from cells derived from the fetus which are floating in this fluid.

Using certain procedures it has been possible to have a child of a preferred sex. The technique involves "sephadex gel column" in which sperms with Y chromosome are trapped in the gel as they are lighter and the X bearing sperms reach the bottom as they are heavier. After separation these sperms are used to fertilize the ova by artificial insemination.

FORENSIC MEDICINE

Following two techniques have been used in the area of forensic medicine to identify the culprits and criminals.

DNA FINGERPRINTING USING MINISATELLITE DNA

The DNA fingerprinting method was developed by Alec Jeffreys and his colleagues in 1985-86. In this technique, DNA is isolated from the blood stains, semen or hair roots and then subjected to southern blotting and DNA hybridization with the help of specific DNA probes. The probes correspond to hypervariable minisatellites in DNA which in turn is made up of tandem repeats of short sequences.

This technique helps in the identification of criminals, rapists and accused using the blood stains, semen stains etc on the victim's clothes or body. It will also help in solving the parental issues in case of doubtful parentage.

Autoantibody Fingerprinting using Dipsticks

Auto antibodies are a novel class of antibodies that react with cellular components in humans and other animal species. These human auto antibodies increase in number from birth up to the age of two years after which they remain constant for decades, if not life long. The complement of these antibodies present in an individual is unique and for this reason they have been named individual specific auto antibodies. These IS auto antibodies when physically separated comprise an antibody finger print that serves to identify people. For this also, body fluids such as blood, semen, tears, saliva and perspiration can be used.

The autoantibody fingerprinting protocol involves:

- Preparation of antigen
- Antibody fingerprint assay

The protocol for autoantibody fingerprinting involves the following steps:

- A panel of antigen is prepared from an extract of human cells in vitro which involves first the separation by denaturing polyacrylamide gel electrophoresis (SDS-PAGE) on the basis of molecular mass.

- Electrophoretic transfer of the antigens to a nylon membrane.
- Blocking of the unbound sites by using the blocking agents.
- Cutting the membrane into strips called Dipsticks.
- Incubation of the dipsticks with dilutions of sera or plasma for 1 hour.
- After washing with buffer, incubate in detector molecule (1 micro-Ci/assay of 125 I protein, which binds to the Fc protein of human IgG in the immune complex formed) for 1 hour.
- After re-washing and drying scan it with a gamma scanner or an optical scanner or perform autoradiography on X-ray film.
- The results are analyzed for the identification of individuals.

Comparison of DNA Fingerprinting and Autoantibody Finger Printing

The advantages of autoantibody fingerprinting over DNA fingerprinting include the following:

- Sensitivity-less than 10 microl blood needed.
- Rapidity-only few hours needed as time consuming method of DNA extraction is not required.
- Simplicity-no requirement for any equipment.
- Cost effectiveness
- Portability
- Autoantibody fingerprints are similar in the newborn child and the mother but are different in genetically identical individuals like identical twins. It is not possible to distinguish this using DNA fingerprinting.

GENE PROBES

Gene probes are short lengths of single-stranded DNA or RNA capable of binding to complementary single-stranded DNA segments which are used to pinpoint the regions on chromosomes where the defective genes are located using recombinant DNA technology. Gene probes have provided a specific test for single-gene disorders which can be used to treat these disorders in the embryo in womb and in young children. Gene probes are also used to detect the presence of DNA from pathogens, as in the diagnosis of AIDS and other viral conditions.

In the future DNA probes may become available for the diagnosis of cancer. Cells become cancerous only after certain genes, called oncogenes, have been altered. Once altered, some cancerous cells produce a specific type of mRNA, not found in healthy cells. It should soon be possible to detect this mRNA by using a probe.

Procedure for the identification of gene fragments with gene probes:

- The fragments are transferred to a well in a flat plate of agarose gel, contained in a box.
- The fragments are separated by electrophoresis. The fragments move

through the gel at different rates, depending on their length. Small fragments travel faster than larger ones.

- After several hours, the fragments become arranged in parallel bands, each band containing many DNA fragments of uniform length.
- The pattern is then plotted onto a nitrocellulose filter.
- Heating binds the DNA to the filter and also separates each DNA fragment into its two single strands of DNA that contain complementary base sequences. Such binding is detected by placing an X-ray film over the nitrocellulose filter.
- When the film is developed, a radioactive region shows up as a dark band.

Using this technique, the DNA sample of sickle cell carrier was studied. Sickle cell anaemia is caused by a mutation in a region of DNA that codes for the protein part of a haemoglobin. One of the bases in DNA molecule, thymine is replaced by adenine. That, in turn, leads to an amino acid sequence in which glutamate is replaced by valine. Due to this defect, the restriction endonulcease M st II cuts the DNA of people with normal haemoglobin (HbA) and sickle cell haemoglobin (HbS) into fragments of different lengths.

GENE THERAPY

A possible cure for single-gene disorders, such as sickle cell anaemia, lies in gene therapy.

Gene Enhancement therapy is an attempt to boost the effect of a gene that is generally thought to be desirable. Height, intelligence and athleticism are traits resulting from complex gene interactions. Once the individual genes that contribute to these attributes have been identified, attempts could be made to add additional copies to the genome. Somatic gene replacement therapy is an attempt to replace defective genes with normal genes.

The Therapy involves adding normal genes to the genome so that the recipient gains a normal gene, while still retaining the defective ones specially in single-gene disorders in which homozygous recessive (aa) individuals show symptoms of the disease.

For example to cure sickle cell anaemia, the process of therapy would involve insertion of a single copy of the normal gene [HbA] followed by multiplication of transformed cells in a suitable growth medium. At an appropriate time, the bone marrow cells of the donor are destroyed using radiation and drugs.

These would be replaced by an injection of the donor's own genetically engineered bone marrow cells. Further treatment would ensure that the engineered cells continued to grow and multiply. At a point where the transplanted cells produced enough haemoglobin [HbA] to meet the individual's needs, a 'cure' would have been affected.

Two important aspects related to gene replacement therapy are: The first is that treated individuals continue to pass on defective genes in their sperm and eggs. The second point is that the treated individuals may carry more number of genes controlling a particular trait and any harmful effects of this combination may take place several years before they begin to show up.

In future, gene therapy could be used for finding a permanent cure for single gene disorder by inserting normal genes in to sperm and eggs. This is Germ-line gene replacement therapy (eugenic gene therapy). This would enable a cure to be passed to the offspring of the treated individuals.

GENETIC FINGERPRINTING

Genetic fingerprinting technique allows the identification of individuals from the lengths of their DNA fragments after samples of their DNAs have been treated with restriction nucleases. This technique has been very effective in resolving the pedigree and/or parentage issues and as diagnostic tool in crime detection.

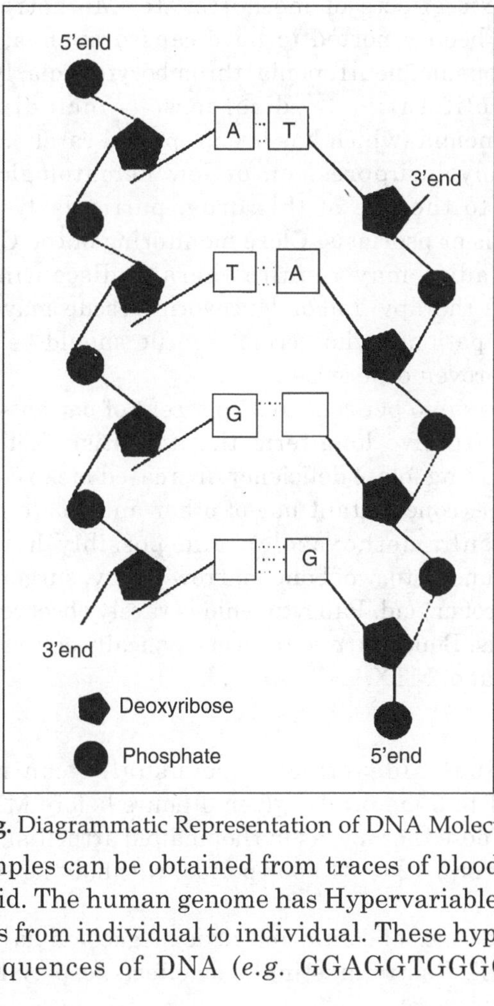

Fig. Diagrammatic Representation of DNA Molecule

The DNA samples can be obtained from traces of blood, hair or semen or any other body fluid. The human genome has Hypervariable regions of various lengths that varies from individual to individual. These hypervariable regions contain short sequences of DNA (*e.g.* GGAGGTGGGCAGGAG) called minisatellites that are repeated many times over. Individual differences in the lengths of restriction fragments appear to depend on, in part, on the number of times each minisatellite is repeated within the hypervariable region.

Using the genetic probes, the DNA fragments from the hypervariable region are separated by electrophoresis and radio-labeled to produce a genetic fingerprint. A completed finger print consist of many parallel bars unique to each individual. The genetic finger print of a child seems to be derived from it's father or mother. Therefore this method is used to establish true blood relationships in cases of disputed parentage.

SIDE EFFECTS BY BODY SYSTEM

Hematologic

Hematologic side effects have included myelosup-pression, which is one of the primary toxic effects of methotrexate. Methotrexate suppressed hematopoiesis has been reported to have caused anemia, aplastic anemia, pancytopenia, leukopenia, neutropenia, thrombocytopenia, lymphadenopathy, and lymphoproliferative disorders including reversible hypogammaglobulinemia (which has been reported rarely).

Pre-existing myelosuppression or low hematologic cell counts are contraindications to the use of this drug, particularly in patients with rheumatoid arthritis or psoriasis. Close monitoring of the CBC is mandatory. Profound count nadirs may require therapy discontinuation, at least temporarily. Folate therapy and/or leucovorin rescue may be preventive or palliative. Treated patients who become febrile should be assumed to have neutropenia until proven otherwise.

Cytopenia occurs in 5 per cent to 25 per cent of patients with rheumatoid arthritis (RA) who receive long-term therapy. Risk factors include renal dysfunction, pre-existing folate deficiency, increased mean corpuscular volume value, advanced age, concomitant use of other anti-folate medications (such as trimethoprim-sulfamethoxazole), and possibly hypoalbum-inemia, concomitant infection, history of bone marrow injury, surgery, and concurrent use of NSAIDs or probenecid. Pancytopenia is rarely observed in patients with rheumatoid arthritis. Bone marrow recovery typically occurs within two weeks after the withdrawal of MTX.

Gastrointestinal

Gastrointestinal side effects are usually controlled by folate supplementation (1 to 5 mg orally, given 4 hours before MTX or 1 mg orally once a day if on low dose therapy, as in rheumatoid arthritis), dosage reduction, dividing the dose over a 12 to 24 hour period, withholding the drug, or giving it parenterally.

Extremely rare cases of colitis and toxic megacolon have been associated with the use of MTX. Gastrointestinal side effects, especially with high-dose administration, may be expected. Serious nausea, vomiting, diarrhea, or stomatitis (10 per cent to 80 per cent of patients followed in long-term studies) may result in symptomatic dehydration. Other frequently reported GI side effects, particularly after high-dose therapy, include gingivitis, pharyngitis, stomatitis, anorexia, hematemesis, melena, gastrointestinal ulceration and bleeding, and enteritis. GI symptoms are often eliminated by folate supplementation. Folate does not affect efficacy of MTX.

Hepatic

Hepatic side effects including hepatotoxicity, acute hepatitis, chronic fibrosis and cirrhosis, decrease in serum albumin, and liver enzyme elevations have been reported.

Methotrexate can cause acute elevations of liver function tests (elevated serum transaminases in 15 per cent of patients with rheumatoid arthritis (RA) on low-dose therapy) or chronic hepatotoxicity (fibrosis and cirrhosis). The incidence of liver fibrosis and cirrhosis in patients with RA (low doses) averages 3 per cent to 7 per cent and 0.1 per cent, respectively.

Meta-analysis has revealed the incidence of progression of liver disease (worsening of 1 grade on the histologic classification of Roenigk) in patients with RA or psoriatic arthritis averages 27 per cent, or 7 per cent per gram of MTX (total dose) given.

Chronic hepatotoxicity typically develops only after chronic use of higher doses (2 years or more of total doses of 1.5 grams or more), is more likely in patients who ingest ethanol, who are aged, who are obese, who have chronic renal insufficiency, or who have diabetes.

In general, the following serve as guidelines for patients with non-oncologic conditions:

- Screen all patients to be treated with complete liver blood tests (transaminases, albumin, alkaline phosphatase, bilirubin), hepatitis B and C serologic studies (some experts recommend viral serology's only in patients with abnormal liver function tests), baseline serum creatinine and complete blood count (CBC);
- Baseline liver biopsy if the patient has pre-existing liver disease, persistently abnormal baseline AST (aspartate aminotransferase), history of excessive alcohol consumption (greater than 3 drinks/day), or chronic hepatitis B or C infection;
- Monitor hepatic enzymes every 4 to 8 weeks (with the understanding that they are not necessarily predictive of fibrosis/cirrhosis), discontinuing MTX if serum transaminase levels meet or exceed 2 to 3x baseline (some guidelines recommend withholding MTX for 1 to 2 weeks if significant abnormalities persist);
- Liver biopsy should be considered if significant hepatic enzyme abnormalities persist for 2 to 3 months. Liver biopsy should be performed if 6 of 12 monthly AST levels are above the upper limit of

normal or serum albumin levels are less than normal in the setting of well controlled disease;
- If liver biopsy shows Roenigk grade I, II, or IIIA (mild fibrosis), MTX may be resumed with monitoring as described above; if liver biopsy show Roenigk grade IIIB (moderate fibrosis) or IV (cirrhosis), MTX should be discontinued;
- MTX should be discontinued in patients with significant liver abnormalities, as described above, who refuse liver biopsy;
- It is still debated whether post-therapy biopsy should be performed every 2 to 3 years or after each 1.5 to 2.0 grams cumulative dose given, regardless of liver chemistries.

*Pre-biopsy coagulation studies and withholding of aspirin and other NSAIDs are recommended.

Immunologic

New and/or opportunistic infection can arise during or after therapy with methotrexate due to drug-induced immune suppression. Infections during MTX therapy may occur in up to 58 per cent of patients during low-dose therapy (as in RA). Patients are at great risk after high-dose therapy. MTX should be held prior to and during elective surgery to minimize the risk of infectious complications.

Limited data have shown that the use of low-dose MTX for patients with rheumatoid arthritis is associated with decreased CD8 and naive CD4 T lymphocytes after only 8 weeks of therapy.Methotrexate is usually contraindicated in patients with overt or laboratory evidence of immunodeficiency syndromes. Immunization may be ineffective when given during methotrexate therapy. Immunization with live virus vaccines is generally not recommended. There have been reports of disseminated vaccinia infections after smallpox immunization in patients receiving methotrexate therapy.

Immunologic side effects including case reports of sometimes-fatal opportunistic infections have been reported. Pneumocystis carinii pneumonia has been reported most frequently. Also of concern are infections associated with severe immunosuppression, such as disseminated herpes zoster, Listeria meningitis, Mycobacterium avium intracellulare pneumonia, and systemic fungal infections (cryptococcosis, nocardiosis, aspergillosis, and histoplasmosis).

Nervous system

Nervous system side effects include headaches, dizziness, drowsiness, blurred vision, subtle cognitive dysfunction, moodiness, tinnitus or unusual cranial sensations. After intrathecal administration, headache, back pain, fever, and even transient paraplegias have been reported. A case of acute chorea and a case of a woman who couldn't speak after receiving methotrexate by intrathecal administration have also been reported. Serious neurotoxicity has been associated with the use of high-dose MTX after intrathecal or intraventricular administration to patients who have undergone craniospinal irradiation, but has also been described in other patients who have received low-dose oral therapy.The development of serious neurotoxicity is more likely if MTX is given in high doses to patients who have undergone craniospinal irradiation.

Significant neurotoxicity can present as an acute stroke-like encephalopathy or a chronic leukoencephalopathy. Unique features of the former include its acute onset (seizures, confusion, hemiparesis, speech problems, loss of consciousness) and reversibility within days.The use of high-dose IV MTX has been associated with the development of chronic delayed leukoencephalopathy in patients with or without a history of craniospinal irradiation.

A syndrome of subtle personality changes followed by a progressive dementia, focal seizures, pseudobulbar palsy, spastic quadriparesis, and stupor may begin several months after the initiation of therapy. Affected patients often improve after discontinuation of therapy.

Intrathecal or intraventricular administration of MTX has been associated with the acute onset of headache, nausea, vomiting, fever, back pain, dizziness, and meningismus. Acute symptoms usually resolve within one to two days. In some cases, subacute neurotoxicity, with myelopathy or encephalopathy, may occur within days to weeks.

Subacute neurotoxicity is usually characterized by paresis/plegia, dementia, confusion, tremor, ataxia, irritability, and somnolence. Delayed leukoencephalopathy, as described above, has also been associated with the use of intrathecal MTX. The condition can improve but can be progressive and fatal.Progressive dementia and leukoencephalopathy has been reported in at least one case after low-dose, oral administration of MTX for the treatment of rheumatoid arthritis.

Respiratory

Two patterns of pulmonary toxicity have been reported a hypersensitivity reaction and a toxic reaction with diffuse alveolar damage and non-specific lung injury. Risk factors are unknown, but may include male gender, cigarette smoking, and concomitant use of an NSAID.

The hypersensitivity reaction is characterized by interstitial pneumonitis, granuloma formation, and the development of bronchopneumonia. Some patients have asthmatic symptoms.

Interstitial pneumonitis may be associated with duration of treatment or weekly or cumulative dose. The prevalence of this complication ranges from 2.5 per cent to 7.5 per cent in patients with psoriatic or rheumatoid arthritis. One prospective study has shown that the cumulative dose of MTX is independently related to small, but significant, increases in residual volume and that MTX may cause mild degrees of air trapping after long-term therapy.

The differential diagnosis of MTX pneumonitis includes infectious pneumonia, hypersensitivity pneumonitis caused by other drugs and autoimmune alveolitis due to rheumatoid arthritis (RA). Pneumocystis carinii pneumonia (PCP) can resemble MTX pneumonitis, and bronchoalveolar lavage (BAL) is a useful tool to distinguish between the two. BAL is not necessarily helpful in distinguishing MTX pneumonitis from RA alveolitis, however, since an elevated lymphocyte count and an increased CD4/CD8 ratio are observed in both conditions.

Pulmonary function tests in affected patients reflect a restrictive ventilatory defect and decreased oxygen-diffusing capacities. A mild peripheral eosinophilia is often present.

Respiratory side effects including toxicity can occur at any dosage in 3 per cent to 11 per cent of patients, and can mimic infectious pneumonia. Interstitial pneumonitis has emerged as one of the most unpredictable and potentially life-threatening side effects from low-dose MTX therapy. It presents as fever, dry/non-productive cough, dyspnea, hypoxemia, and/or epistaxis, usually with chest X-ray infiltration. Signs or symptoms of infection must be treated seriously due to the possibility of MTX-induced neutropenia.

Renal

Renal side effects include renal insufficiency, which is most commonly associated with high-dose MTX since, after these doses, the concentration of a major circulating metabolite, 7-OH MTX, can precipitate in the renal tubule. Concomitant use of other potentially nephrotoxic drugs, including NSAIDs, and pre-existing renal insufficiency are risk factors. There is a higher risk of nephrotoxicity in patients with underlying renal dysfunction. Aggressive and adequate hydration and urinary alkalinization helps minimize the risk of MTX-induced nephropathy, cystitis, and hematuria. If monitoring reveals renal dysfunction, decreasing MTX doses or discontinuing the drug altogether may improve renal function.

General

General side effects, which have been reported frequently, include malaise, fatigue, and chills. Less commonly reported effects include arthralgias and myalgias. Leucovorin is typically given to diminish toxicity and counteract the adverse effects of high-dose therapy. There is an increased risk of methotrexate (MTX) toxicity in patients who have "third space" accumulations of fluid, such as pleural effusions. MTX accumulates in these spaces and becomes cleared from these spaces slowly. These "deep pools" of MTX markedly increase the risk of toxicity, particularly gastrointestinal toxicity (mucositis) and should be evacuated before therapy begins.

Genitourinary

Genitourinary side effects may seriously affect either sex. Women may experience menstrual dysfunction, vaginal discharge, abortion, or infertility. Decreased libido has been described in each sex. Defective oogenesis and spermato-genesis is usually transient.

Dermatologic

Cases of severe, sometimes fatal, dermatologic reactions, such as toxic epidermal necrolysis, Stevens-Johnson syndrome, exfoliative dermatitis, skin necrosis, and erythema multiforme, have been reported in children and adults, within days of oral, intramuscular, intravenous, or intrathecal methotrexate administration.

Reactions have been observed after single or multiple, low, intermediate or high doses of methotrexate for the treatment of neo-plastic and non-neo-plastic diseases. Alopecia typically resolves several months after discontinuation.

Dermatologic side effects include erythematous rashes, desquamation, epidermal necrosis, pruritus, urticaria, photosensitivity, pigmentary changes, alopecia, ecchymosis, telangiectasia, acne, cutaneous vasculitis, furunculosis, and alopecia. Painful plaque erosions have rarely been reported (when MTX was used to treat psoriasis).

Musculoskeletal

Musculoskeletal side effects from low-dose MTX in patients with rheumatoid arthritis include rare instances of accelerated nodulosis. This problem is more commonly associated with MTX than other second-line agents. The use of MTX may be associated with decreased bone density in a syndrome called "MTX osteopathy" (this potential complication requires further investigation). This syndrome may be especially important in postmenopausal women.

Hypersensitivity

Hypersensitivity side effects including anaphylaxis have been reported rarely.

Oncologic

Oncologic side effects include some evidence that methotrexate may be oncogenic, particularly with respect to the development of some lymphomas and leukemias.

Large retrospective studies have shown that hematologic malignancies are uncommon in patients with RA treated with disease-modifying antirheumatic drugs, including MTX. These studies have shown that there does not appear to be a relationship between the peak or cumulative dose of the duration of MTX therapy and the subsequent development of hematologic malignancy. The histologic types of hematologic malignancies seen in MTX-treated patients do not appear different from those seen in patients with RA treated with other disease-modifying antirheumatic drugs. Underlying rheumatoid arthritis (RA) or Sjogren's syndrome are independent risk factors for the development of non-Hodgkin's lymphoma.

Cardiovascular

Cardiovascular side effects including pericarditis, pericardial effusion, myocardial ischemia, hypotension and ventricular arrhythmias have rarely been associated with MTX. Thromboembolic events including arterial thrombosis, cerebral thrombosis, deep vein thrombosis, retinal vein thrombosis, thrombophlebitis, and pulmonary embolus have also been reported with methotrexate use. Chemical pleuritis secondary to MTX has been identified as the cause of chest pain in some patients after high-dose therapy.

Endocrine

Endocrinologic side effects have included gynecomastia associated with the use of low-dose MTX in patients with rheumatoid arthritis.

Ocular

Ocular side effects reported with methotrexate include conjunctivitis and serious visual changes of unknown origin.

PHARMACOGENOMICS

The term pharmacogenomics (or pharmacogenetics, the two terms can be used interchangeably) is derived from pharmacology (study of pharmaceuticals) and genetics, so it is the study of how a person's body reacts to pharmaceuticals, given that person's specific genetic make-up. Widespread application of pharmacogenetics is not done at present, but medical scientists believe that it has great potential to improve current therapies.By knowing an individual's genetic profile, a doctor would be able to prescribe the correct medication, at the correct dosage.

The risk of adverse reactions, side effects and overdosage would therefore be minimal.

The basis of pharmacogenomics is the identification of SNPs (pronounced as 'snips', derived from the abbreviation for 'single-nucleotide polymorphisms'). SNPs are differences between individual human beings of a single base pair in their DNA.In the past, the sequencing of a person's DNA was a lengthy and expensive procedure, but with the development of the DNA microarray (or DNA chip, as it is also called) the sequencing can be done quickly. SNPs can be used to map and identify specific genes that play a role in diseases such as diabetes, cancer and arthritis.

The proteins that these genes encode for can become targets for new therapies. As such, pharmacogenomics can play an important role in oncology, treatment of high blood cholesterol levels, tailoring treatment for people with psychiatric disorders, treatment for people with cardiovascular diseases.

However, the use of pharmacogenomics to 'individualise' treatment is still in its infancy. Pharmacogenetic testing to determine an individual's possible reaction against treatment is currently most advanced in Scandinavian countries. Here it is mainly used for CYP2D6 genotyping(this word used for the first time), which aids individual dose selection to treat psychiatric illness.

The Role of Biotechnology in Developing Vaccines

The word 'vaccine' was derived from the Latin for 'cow' (*vacca*)–referring to Edward Jenner's discovery in 1796 that milkmaids, who were in frequent contact with cowpox were immune against the dreaded smallpox. A vaccine is a harmless biological preparation that is given to humans to make them immune against a specific disease.

The human body's immune system recognises the vaccine as being 'foreign', destroys it, but also 'remembers' what this foreign matter looked like. When the body then actually encounters the 'real' disease (or virulent form), the immune system recognises it and will be ready to fight off the infection. Scientists may take one of several routes to develop a vaccine, depending on how the disease-causing microbe infects body cells, how the body's immune system reacts, physical characteristics of the microbe and also where the vaccine is going to be used. The various approaches are the following:

Live, Attenuated Vaccines

These vaccines contain a version of the disease-causing microbe that has been weakened (attenuated), so that it cannot cause disease but only prompt the immune system to remember it. Live attenuated vaccines cause a very strong immune reaction, so that only one or two doses generally give lifelong immunity.

It is mostly used against viral diseases, such as measles, mumps and chickenpox. It is not safe to use a live attenuated vaccine on a person with a weakened immune system (HIV-positive individuals or patients receiving chemotherapy). These vaccines also need to be refrigerated to stay potent, limiting their use in some developing countries. There is also the remote possibility that the weakened microbe might mutate back to its virulent form and cause disease.

Inactivated Vaccines

In this case, the disease-causing microbe is killed with chemicals, heat or radiation, and not just weakened. The microbe can therefore not mutate back to its virulent form.

However, this type of vaccine does not produce such a strong immune reaction, so additional immunisation ('booster shots') are necessary. Inactivated vaccines are freeze-dried, so they can be stored easily–making them better for use in developing countries. Examples of inactivated vaccines are those against cholera, bubonic plague and hepatitis A.

Subunit Vaccines

These vaccines do not include the entire disease-causing microbe, but only the antigens that stimulate the immune system the most. Antigens are 'markers' on the surface of a microbe, and this is the part that is recognised by the immune system's T-cells, and to which the T-cells bind. Scientists can make the antigens from the microbes in the laboratory using recombinant DNA technology. In this case the vaccine is called a recombinant subunit vaccine, such as the vaccine against the hepatitis B virus.

Toxoid Vaccines

These are vaccines that are used against bacteria that secrete toxins, for example diphtheria and tetanus. A toxoid vaccine is made by treating the toxin with formalin, rendering the toxin harmless. The vaccine causes the immune system to produce antibodies against the toxin, which bind to the toxin and block its action.

Conjugate Vaccines

Many harmful bacteria have an outer coating of sugar molecules known as polysaccharides. This coating hides the antigens (markers) on the surface of the bacteria so that the immature immune system of a child or baby cannot recognise it. Scientists link antigens or toxoids from a microbe that an immature immune system can recognise to the polysaccharides, thereby making a conjugate vaccine. The vaccine against Haemophilus influenzae type B (Hib) is a conjugate vaccine.

DNA Vaccines

These vaccines are still in the experimental phase, but several types are already being tested in humans. A DNA vaccine only uses the genes of the microbe that code for the antigens of that microbe. When those genes enter the body, they are taken up by body cells, and then instruct the body cells to produce antigens. The antigens then stimulate the immune response. DNA vaccines are easy to produce and store. Vaccines against herpes and influenza are currently being tested.

Recombinant Vector Vaccines

Recombinant vector vaccines are similar to DNA vaccines, but they use an attenuated virus or bacterium as a vector to carry the DNA of the disease-causing microbe into the body. The vector virus then 'infects' body cells, thereby delivering the DNA to the body cells. Researchers are working on viral-based and bacterium-based recombinant vector vaccines against HIV, rabies and measles. A vaccine can be monovalent (immunizes against one disease), or multivalent (immunizes against more than one disease, or two or more strains of the same disease-causing microbe).

Regulation and Legislation of Biotechnology

The regulatory process ruling the use and development of biotechnology started soon after Boyer and Cohen's discovery of the recombinant DNA technique. In mid-1974, scientists called for a voluntary moratorium on certain experiments with recombinant DNA.This was followed by the Asilomar Conference, during which scientists from all over the world, lawyers and government officials debated the way forward.

The conference concluded that research on recombinant DNA should be ruled by strict guidelines, which were then issued by the National Institutes of Health.According to these guidelines, all experiments and trials concerning human gene transfer ('gene therapy trials') making use of recombinant DNA technology must be reviewed by the Recombinant DNA Advisory Committee of the NIH.

Researchers who receive funding from the NIH for their work (many South African scientists receive funding from this source), or who conduct their research at facilities receiving funding from the NIH, are bound by these guidelines. However, researchers who receive private funding, and who conduct their research at privately funded institutions, are not bound by the guidelines.

Locally, research institutions such as the Medical Research Council has an Ethics Committee that is registered with the Office for Human Research Protection in the USA.The mandate of the MRC's Ethics Committee is to review all applications for funding of medical research to ensure that the goals of the project do not violate the sanctity of life and obey all rules. The Committee has developed a set of research guidelines on various forms of research, including human genetic research. In addition, every tertiary institution where such research is conducted, also has its own ethics committee that oversees research experiments.

South Africa has no specific law that governs the biotechnology industry or research field, but according to Prof. Tony Bunn and Dr. Michelle Mulder of the MRC's Innovation Centre, has legislation that governs different aspects of the subject.

The Department of Science and Technology published a National Biotechnology Strategy in 2001, which outlines the government's plans to build the biotechnology industry in South Africa. Legislation mainly covers safety and ethical issues; as well as intellectual property rights. Safety and ethical issues are dealt with in the National Health Act No 61 of 2003(Department of Health), which contains a chapter on the use of blood, blood products, tissue and human reproductive cells (sperm and ova) during medical research.

The Act also addresses the issue of human cloning. Intellectual property rights are covered from two fronts. The Companies and Intellectual Properties Registration Office (CIPLO), which forms part of the South African Department of Trade and Industry has laws that are applicable to biotechnology (for example Section 25 of the Patent Act, Act 57 of 1978applies to scientific discoveries).

The Department of Science and Technology's Intellectual Property Rights from Publicly Financed Research and Development Billdeals with scientific discoveries, for example gene therapies, that are made at publicly financed institutions, such as universities.

- There is still no cure for many of the diseases that can be diagnosed or predicted by means of biotechnology. Knowing that you carry a gene for a future untreatable disease might place a psychological burden on an individual. Also, genetic tests cannot always accurately predict future disease. Parents of unborn children could face a difficult choice about ending a pregnancy, or not.
- The ownership of genetic information–who will own the information gleaned from a genetic test? Diseases that are found only among certain groups (for example Tay-Sach's disease that is common to people from Jewish East European descent) can lead to stigmatisation and discrimination. The United States Senate has recently unanimously passed legislation that will only allow patients and their doctors access to data from genetic testing, thereby banning genetic discrimination.
- Any error made when germline therapy is carried out, will be passed on to future generations. Although this therapy is not being performed yet, many have raised concerns that this might change the basic nature of human beings by altering their genetic make-up.
- Pharmacogenomics will make the development and testing of new pharmaceuticals cheaper, because participants to clinical trials could be pre-screened to show if the pharmaceutical being tested would be harmful or ineffective to them. This will make the trials smaller, faster and therefore cheaper.
- Economics play a vital role in the development of new treatments. Three important diseases against which a vaccine would be of immense benefit, HIV malaria and tuberculosis, exist mainly in poor countries. Biotechnology and pharmaceutical companies have little incentive to develop vaccines, hence there is minimal financial return.
- Despite the Human Genome Project, we still only have a limited knowledge of the functions of genes. It is not known exactly whether genes have more than one function. So if genes are replaced during gene therapy, this might influence other body processes.
- Most genetic disorders involve more than one gene, as well as interaction with the environment. Diet, lifestyle and other environmental factors play an important role. For example, genetic tests can show whether a woman carries a gene mutation at position BRCA1 (BRCA1 is the name of the gene not a location: rewrite it as follows: "For example, genetic tests can show whether a woman carriers a mutation in the BRCA1 gene". This puts her at risk for breast cancer. But not everybody that carries this mutation develops

breast cancer. Conversely, if genetic tests reveal that a person, for example, does not have a gene putting him at risk for cardiovascular disease, this might lead to carelessness.

THE ROLE OF BIOTECHNOLOGY IN DIAGNOSTICS

Biotechnology plays a vital role in modern diagnostic science. Through our current knowledge of DNA and genes, we can examine a person's DNA (usually taken from a blood sample) to do the following:

- Determine the sex of an unborn baby
- *Carrier screening*: The identification of unaffected individuals who carry one copy of a gene for a disease that requires two copies of a gene to manifest, for example haemophilia, or Tay-Sachs disease (disorder of the body's ability to metabolise fats)
- *Prenatal diagnostic screening*: For example screening for Down syndrome
- *Newborn screening*: Testing newborn babies for HIV, phenylketonuria (an inborn metabolic disorder, leading to abnormally high levels of the amino acid phenylalanine, which if left untreated can cause severe mental retardation)
- *Presymptomatic testing for predicting adultonset disorders*: For example familial high blood cholesterol (GeneCare, a Cape Town based company designed a comprehensive genetic test for cardiovascular disease, incorporating other risk factors such as lipid metabolism, blood clotting, folate/homocystein metabolism, and iron metabolism
- Presymptomatic testing for estimating the risk of developing adult-onset cancers (for example colon cancer and bladder cancer). Persons with a mutation in the BRCA1 gene has a 65 per cent cumulative chance of developing breast cancer
- *Confirmational diagnosis of symptomatic individuals*: For example cystic fibrosis and Huntington's disease
- *Forensic/identity testing*: For example analysis of semen for prosecution of sexual offenders, analysis of blood, bone and hair of murder victims, paternity testing. During diagnostic DNA testing, scientists scan patients' DNA for errors, also known as mutations. These mutations can be large (for example a piece of chromosome missing, or added) or small (a difference in one of the 'building blocks' or chemical bases making up the double helix of DNA). Sometimes pieces of chromosomes can become 'switched', so that genes end up in the wrong positions on the chromosome.Two techniques are mainly used. In the first, the scientist uses short pieces of DNA (which are called 'probes'), that have sequences complementary to the suspected mutations. These probes seek the complementary (if this is for public consumption then "complementary" means nothing in DNA context)

sequences among the patient's DNA, bind to it and mark or flag it. The second technique involves comparing the sequence of bases in the patient's gene to that of a healthy individual.

The Role of Biotechnology in Therapeutics

Biotechnology contributes to the development of treatments against disease in two ways: gene therapy and pharmacogenomics.

GENE THERAPY

Although gene therapy is still in its infancy, there have been cases where patients have been treated with this technique. Essentially, gene therapy involves the treatment of disease by changing the genetic message or instructions of body cells.

Gene therapy can follow one of three approaches: the replacement of a faulty gene with a normal gene; the inactivation or 'knocking out' of the faulty gene; or introducing a completely new gene into the body to fight the disease.At the moment, only somatic gene therapy is performed. This is gene therapy involving the body cells of patients. Any change in genetic instructions remains within this individual and will not be passed on to his or her children. Germline therapy is gene therapy that involves the reproductive cells (sperm in men, and ova in women).

Changes in genetic instructions will be passed on to the patient's children. However, at this stage, this type of gene therapy is still experimental and it is not being performed on any patients. Scientists use a carrier, or vector, to transport a gene into the cells of a patient. Viruses are often used as vectors, but their disease-causing genes are first removed so that they cannot cause disease. The vector virus carrying the new gene can be injected into the patient, or given intravenously.

The vector then 'infects' the target cells and delivers the new gene into the cell. The first case of gene therapy being used was reported in 1990, when doctors at the National Institutes of Health (USA) treated a toddler girl suffering from severe combined immune deficiency (SCID).

The movie The boy in the bubble tells the story of David Vetter who suffered from the same disease. SCID is caused by an abnormal ADA gene, which is the gene that regulates the production of an enzyme, adenosine deaminase.

This enzyme is vital to the normal functioning of the immune system. The doctors removed bone marrow cells from the little girl, treated them with a vector carrying a normal ADA gene, and then returned the treated bone marrow cells to the little girl.After the treatment, her immune system started to function normally.

Because a faulty gene is also the cause of cystic fibrosis, the approach of replacing the faulty gene with a normal, healthy one, could also be used to treat this disease.

A phase I clinical trial is currently being conducted on gene therapy against cystic fibrosis. Because scientists were not convinced of the efficacy of the therapy, it is now being reformulated before further testing takes place.Scientists are also investigating how gene therapy can be used against cancer.Various possibilities are being investigated, ranging from replacing missing or altered genes that can cause cancer, to introducing new genes into cancer cells, making them more vulnerable to treatment.

7

Genetics Revolution in Biotechnology

A BIOTECHNOLOGICAL REVOLUTION

The impact of technological innovation may be measured by the depth and breadth of its penetration into the industrial economy. The depth of its penetration is determined by how great a transformation the technology makes within a single sector. Thus, if advances in photovoltaic technology were to result in solar cells that generate 50 per cent of the domestic electricity, this advance would be considered a "deep" revolution in energy production.

The breadth of a technological revolution signifies how many sectors of the economy are affected. In 1908, L. H. Baekeland invented a synthetic substance called bakelite that ushered in the age of plastics. Advances in material sciences such as this have affected every sector of the industrial and consumer economy and as such represent a technological revolution of considerable penetration.

By most accounts, several discoveries that were made in the biomedical sciences during the late 1960s and early 1970s have spawned a major technological revolution popularly referred to as biotechnology. Unlike many of the technological innovations introduced during the American Industrial Revolution, which was not a revolution in science, but rather one that was built on practical invention, the biotechnological revolution derives directly and immediately from the pure sciences.

Once genes were capable of being excised, spliced, and transplanted into foreign cells, the power of the technique as an industrial tool was patently obvious to those who played a central role in its discovery. But the pioneers in molecular biology were not seeking an industrial revolution. Far from it, they stumbled on to one while in hot pursuit of new methods to study the genetics of higher organisms.

Several terms have been used to describe the changes in science and technology emanating from the biological research laboratories, including biotechnology, bioengineering, genetic technology, genetic engineering, recombinant DNA, and some more esoteric names like algeny and plasmid engineering. However, many of these terms do not arise *ex nihilo* but have an ancestry both in the processes they describe and in their etymology. The term *biotechnic* appears as a fleeting thought in Lewis Mumford's extensive treatise on culture and machines titled *Technics and Civilization*. In this work Mumford examines the development of technology through a millenium of modern history. The framework he presents for conceptualizing technological change is a useful guide for positioning biotechnology in the historical landscape of major industrial revolutions.

Building on a typology derived from Patrick Geddes, Mumford posits three overlapping and interpenetrating periods of technological transition. Eotechnics represents the period defined by the handicraft industry, where human production was based on the use of natural materials available on the earth's surface, such as water and wood.

Paleotechnics refers to the stage of mass production where subterranean materials, such as coal and iron, were appropriated. Finally, the third stage, which Mumford calls neo-technics, involves the development of electric energy, alloys, synthetics, mass communication, and the application of the concepts of science "to every phase of human experience and every manifestation of life."

Each stage is associated with a period of human history and a family of technologies that share similar energy sources and correlate with a unique material basis and mode of production. By postulating these stages, Mumford can highlight the broad outlines of technological change in civilization. Each phase... has its origin in certain definite regions and tends to employ certain special resources and raw materials. Each phase has its special forms of production. Finally, each phase brings into existence particular types of workers, trains them in particular ways, develops certain aptitudes and discourages others, and draws upon and further develops certain aspects of the social heritage.

According to Mumford, the early stages of industriali-zation subjugated life processes to the machine. But during the neo-technic phase, the objects of mechanization began to be influenced by organic forms. This means that the animate world was beginning to serve as a model for technological design. The ear, the eye, wings, and vocal cords all became important symbols and even templates of mechanical invention. "Machines, which had assumed their own characteristic shapes in developing independent of organic forms, were now forced to recognize the superior economy of nature."

Mumford not only brought attention to the metamorphosis of technics and its interrelationships with culture, but he anticipated human factors research. Instead of humans adapting to a fixed and ponderous technology, he saw instances in which the forms of biological life guided the purposes and structure of mechanical invention. For Mumford, this was an optimistic sign and one of the few counterforces to the rise of technological devices with dehumanizing capacities.

One of the principles of organic life is functionalism. Animals possess what they need to function and survive in some segment of nature. What lacks function, slowly disappears from the species. Writing in the 1930s, Mumford cites the trend in technology design of stripping the machine to its essential functions.

It is a first step towards that completer integration of the machine with human needs and desires which is the mark of the neo-technic phase, and will be even more the mark of the biotechnic period, already viable over the edge of the horizon.

He introduces the term biotechnics to describe a hopeful trend whereby technology, "instead of benefiting by abstraction from life, will benefit even more greatly by its integration with it." Mumford brought attention to the possibilities of technology for becoming life-affirming in its function, its aesthetic, and its physical coexistence with humans.

The term *biotechnics* is a hybrid concept derived from the terms "biology" and "technology." When Mumford introduced the term he was searching for a way to describe how living things can guide technological progress. Fifty years later, the world is a different place. Science and technology are now not so much guided by, as they are guiding, the research and invention of living forms. Life forms have become the substrate of a new technology driven by a mechanistic paradigm. Instead of living things standing in opposition to mechanical invention, as Mumford Mumford viewed them, they have become invention.

Within genes lies the essence of living forms, the *élan vital,* according to the current scientific fashion.

This perspective is supported by the following popular, albeit exaggerated, scientific claims. Genes direct the production of biological matter. They determine the physical characteristics of the life form. They are the basis of all change in organic life through the process of mutation. Increasingly, new genes are being discovered that are cited as a principal factor in the etiology of a disease. And finally, some scientists are pursuing sociobiological theories that posit a genetic basis for social and psychological behaviour in humans.

An industrial revolution invariably influences the way we think about the world. Many new scientific concepts eventually filter into the popular vernacular. The changes in language occur through two processes. First, the instruments of technology arising from scientific discovery establish new relationships among people and between people and nature. These new relationships are soon reflected in the creation of symbols and images for comprehending and describing the social and physical elements in our environment. Second, the ideology of the industrial revolution is transmitted through the literature of the intelligentsia into the popular culture (novels, mass media, and lay renditions of science).

PRINCIPLES OF GENETICS

The site where genes work is the cell. Some organisms, such as paramecia or amoebas, are made up of a single cell. Other organisms are made of many kinds of cells, each having a different function. For instance, a tree contains some cells that form the root system and other cells that form leaves. Each cell's function within an organism is determined by the genetic information encoded in DNA.

In animals, plants, and other eukaryotes (organisms whose cells contain a nucleus), DNA resides within membrane-bound structures in the cell. These structures include the nucleus, the energy-producing mitochondria, and, in plants, the chloroplasts (structures where photosynthesis takes place). In prokaryotes, one-celled organisms and bacteria that lack internal membrane-bound structures, DNA floats freely within the cell body.

Cell Division and Reproduction

Organisms could not grow or function properly if the genetic information encoded in DNA was not passed from cell to cell. DNA is packaged into structures called chromosomes within a cell. Every chromosome in a cell contains many genes, and each gene is located at a particular site, or locus, on the chromosome.

Chromosomes vary in size and shape and usually occur in matched pairs called homologues. The number of homologous chromosomes in a cell depends upon the organism—for example, most cells in the human body contain 23 pairs of chromosomes, while most cells of the fruit fly *Drosophila* contain 4 pairs.

Within all organisms, cells divide to produce new cells, each of which requires the genetic information found in DNA. Yet simply splitting the DNA of a dividing cell between two new cells would lead to disaster—the two new cells would have different instructions and each subsequent generation of cells would have less and less genetic information to work with.

Imagine how chaotic it would be to rip an architectural blueprint in two, give each half to different contractors, and tell them to construct identical buildings.

Just as each contractor would require a full copy of the blueprint to construct a complete building, each new cell needs a complete copy of an organism's genetic information to function properly.

Organisms use two types of cell division to ensure that DNA is passed down from cell to cell during reproduction. Simple one-celled organisms and other organisms that reproduce asexually—that is, without the joining of cells from two different organisms—reproduce by a process called mitosis. During mitosis a cell doubles its DNA before dividing into two cells and distributing the DNA evenly to each resulting cell. Organisms that reproduce sexually use a different type of cell division.

These organisms produce special cells called gametes, or egg and sperm. In the cell division known as meiosis, the chromosomes in a gamete cell are reduced by half. During sexual reproduction, an egg and sperm unite to form a zygote, in which the full number of chromosomes is restored.

Mitosis

Mitosis occurs in five stages: Prophase, prometaphase, metaphase, anaphase, and telophase. During prophase, the start of mitosis, the DNA of each chromosome replicates. Each chromosome then reorganizes into paired structures called sister chromatids, with each member of the pair containing a full copy of the DNA sequence.

The sister chromatids condense, thickening until they appear joined at a single site, known as the centromere. Prometaphase is marked by the disintegration of the nuclear membrane. The sister chromatids line up in the middle of the cell halfway between the poles. In metaphase, exactly half of the chromatids face one pole, and half face the other pole. This equilibrium position is called the metaphase plate. In anaphase, the chromatid pairs split apart at the centromere, and each half of the pair then moves towards opposite poles of the cells. In telophase, the final stage of mitosis, a nuclear membrane forms around the chromosomes at each pole of the cell. Mitosis ends with the formation of two new cells, each with a matching full set of chromosomes as well as an identical complement of cellular structures.

Meiosis

During meiosis, two cell divisions occur to produce four daughter cells from the original parent cell. Each resulting cell has half the chromosomal DNA of the parent cell. A half set of chromosomes in an organism is known as the haploid number. In the first cell division of meiosis the chromosomes of a gamete cell duplicate and join in pairs. The paired chromosomes align at the equator of the cell, and then separate and move to opposite poles in the cell. The cell then splits to form two daughter cells. As meiosis proceeds, the two daughter cells undergo another cell division to form four cells, each of which bears half of the number of chromosomes found in the other cells of the organism.

Meiosis ensures that reproduction will produce a zygote that has received one set of chromosomes from the male parent and one set of chromosomes from the female parent to form a full set of chromosomes. The entire set of chromosomes in an organism is known as the diploid number. Once formed, the zygote continues to divide and grow through the process of mitosis.

Patterns of Inheritance

In life forms that reproduce asexually, such as bacteria and amoebas, all offspring share the exact same genes and are identical to their parents. The genetic transmission that occurs in organisms that reproduce sexually is far more complex. An individual that forms by the union of two gametes inherits its chromosomes from two distinct parents. Consequently, sexual reproduction guarantees that offspring with new combinations of genes will continually arise.

Certain patterns of inheritance were evident long before scientists discovered the molecular structure of DNA and chromosomes. Throughout history, people have recognized that certain traits, whether in humans, animals, or agricultural crops, could be passed from generation to generation. Yet for centuries, people were unable to reconcile many confusing observations about the mechanisms of inheritance.

The first person to make sense of this complex subject was Austrian monk Gregor Mendel, who conducted a series of experiments on pea plants beginning in the 1850s. Mendel observed the results of crossbreeding plants with different characteristics, such as height, flower colour, and seed shape. His conclusions from these experiments led him to develop explanations for how traits are transmitted from generation to generation. Mendel's theories form the foundation of modern genetics.

Mendel's Rules

In his research, Mendel observed that characteristics were inherited as separate units, each of which was inherited independently of the others. Mendel suggested that each parent has pairs of these units but contributes only one of each pair to offspring. The units that Mendel described were later given the name *genes*. Mendel recognized that a gene can exist in different forms.

Today these alternate forms are known as alleles. For example, pea seeds, the edible part of the plant we call peas, have a texture trait controlled by a single gene. This gene occurs in two alleles: one corresponding to round (smooth) peas, the other to wrinkled peas. Although an individual can carry only two alleles for a particular gene, each gene may have dozens of different alleles.

Mendel's experiments focused on interbreeding different strains of pea plants and then observing the traits that appeared in subsequent generations. When he crossbred plants with round peas and those with wrinkled peas, he discovered that all of the resulting offspring had round peas.

Today we know that peas are round and smooth when they contain the right amount of sugar. If peas are missing the gene that produces a protein called starch branching enzyme 1 (SBE1), the peas make too much sugar, causing the peas to swell and then wrinkle and shrivel as they dry. Mendel concluded that when an organism has two different alleles corresponding to the same genetic trait, one of the two may be dominant.

The other allele is said to be recessive, meaning that its presence will be detectable only if an organism has inherited the recessive gene from both parents. For convenience, geneticists designate alleles by a single letter—the dominant allele is represented by a capital letter and the recessive allele by a small letter. In the pea texture example, a plant inherits one allele for pea texture from each parent. The dominant allele that produces SBE1, resulting in round, smooth peas, is designated as R, while the recessive allele that does not produce SBE1 and produces wrinkled peas is designated as r.

To determine the set of alleles an organism has for a given trait just by visual observation can often be difficult. In the pea plant example, for instance, plants with smooth peas might be carrying two dominant alleles for that characteristic (*RR*) or one dominant and one recessive allele (*Rr*). Geneticists use the term *genotype* to refer to the combination of genes that code for a trait, while the term *phenotype* describes the physical manifestation of that trait. Therefore, the presence of two dominant alleles for pea texture (*RR*) would reflect the genotype while a smooth pea indicates the phenotype.

Mendel did not limit his experiments to testing the rules of inheritance of single traits. He also studied plant traits involving multiple pairs of genes, breeding plants that have round, yellow seeds with plants that produce wrinkled, green seeds. Such experiments demonstrated that the patterns of inheritance he observed in his experiments with single traits also apply to cases involving more complex gene combinations.

Exceptions to Mendel's Rules

Mendel published his studies in a science journal in 1865, at which time no other scientist commented on his work. Since that time, geneticists have learned that sometimes genes do not easily conform to so-called Mendelian patterns of inheritance.

Incomplete Dominance

In cases of incomplete dominance, the inheritance of a dominant and a recessive allele results in a blending of traits to produce intermediate characteristics. For example, four-o'clock paint plants may have red, white, or pink flowers. Plants with red flowers have two copies of the dominant allele *R* for red flower colour (*RR*). Plants with white flowers have two copies of the recessive allele *r* for white flower colour (*rr*). Pink flowers result in plants with one copy of each allele (*Rr*), with each allele contributing to a blending of colours.

Quantitative Inheritance

Mendel focused his studies on traits determined by a single pair of genes, and the resulting phenotype was easy to distinguish. A tall plant can be markedly different from a short one, and a green pea can easily be distinguished from a yellow one.

There are some traits, however, that are not easy to distinguish. Human skin colour, for example, may be any of a wide variety of shades. Traits such as skin colour differ from the ones Mendel studied because they are determined by more than one pair of genes.

In this form of inheritance, known as quantitative inheritance, each pair of genes has only a slight effect on the trait, while the cumulative effect of all the genes determines the physical characteristics of the trait. At least four pairs of genes control human skin colour. Multiple genes also control many traits important in agriculture, such as milk production in cows and ear length in corn.

Multiple Alleles

Another exception to Mendelian genetics involves genes with multiple alleles. Certain traits are controlled by multiple alleles that have complex rules of dominance. In humans, for example, the gene for blood type has three alleles: I_A, I_B, and i. With three alternatives for each member of a gene pair, there are six possible combinations of these genes (I_AI_A, I_BI_B, ii, I_Ai, I_Bi, I_AI_B). Although there are six possible combinations, humans have only four major blood types: A, B, AB, and O. This results because both I_A and I_B dominate over i, but not over each other, so a person with a gene combination of I_AI_A or I_Ai has blood type A. The gene combinations I_BI_B and I_Bi both produce blood type B. I_AI_B results in a blood type AB, and ii results in blood type O.

Gene Linkage

In his experiments, Mendel was careful to study traits in pea plants where one trait did not appear to influence another, such as the plant's height or the pea's texture. These two phenotypes (height and texture) occur randomly with respect to one another in a manner known as independent assortment. Today scientists understand that independent assortment occurs when the genes affecting the phenotypes are found on different chromosomes. An exception to independent assortment develops when genes appear near one another on the same chromosome. When genes occur on the same chromosome, they are inherited as a single unit.

Genes inherited in this way are said to be linked. For example, in fruit flies the genes affecting eye colour and wing length are inherited together because they appear on the same chromosome. But in many cases, genes on the same chromosome that are inherited together produce offspring with unexpected allele combinations. This results from a process called crossing over. Sometimes at the beginning of meiosis, a chromosome pair (made up of a chromosome from the mother and a chromosome from the father) may intertwine and exchange sections of chromosome. The pair then breaks apart to form two chromosomes with a new combination of genes that differs from the combination supplied by the parents. Through this process of recombining genes, organisms can produce offspring with new combinations of maternal and paternal traits that may contribute to or enhance survival.

Sex-Linked Traits

Most chromosome pairs consist of identical, or homologous, partners. In many species, including humans, there is one pair of chromosomes in which the partners noticeably differ from each other. These are called the sex chromosomes because they determine the differences between males and females. Genes located on the sex chromosomes display different patterns of inheritance than genes located on other chromosomes. In human females, the sex chromosomes consist of two X chromosomes, while males have an X chromosome and a shorter Y chromosome with many fewer genes. In males the X chromosome contains many genes that have no corresponding gene on the Y chromosome.

A male's X chromosome may contain a recessive allele associated with a genetic disorder, such as hemophilia or Duchenne muscular dystrophy. In this case, males do not have a normal second copy of the gene on the Y chromosome to mask the effects of the recessive gene, and disease typically results. Additional examples of sex-linked traits include red-green colour blindness in humans and eye colour in fruit flies.

GENETICS AND THE ENVIRONMENTALISTS

In the mid-1970s, when rDNA molecule research was still an exotic technique, the public entry into the debate over its safe uses was determined by several factors. First and foremost, a group of distinguished biologists called attention to the new technique by publishing letters in science journals that caught the attention of the popular press. The international meeting at Asilomar was attended by science writers, whose coverage of the events dramatized the disagreements among biologists at the meeting.

In drawing attention to the new discovery, the Asilomar organizers did not intend for the fate of the new technology to be decided by popular acclamation or democratic process. However, some biologists, frustrated by the NIH's role in setting safety standards, brought their concerns to university campuses and local communities. To the public, it appeared that scientists were polarized on the dangers of rDNA research.

Second, the issues were dramatized through the world press by the controversy that erupted in Cambridge, Massachusetts, which pitted community values against the interests of national science. The 1976 Cambridge rDNA controversy represented the birth of public involvement in genetic engineering. Prior to that, the issues were debated exclusively in professional groups and on university campuses, notably the University of Michigan.

Third, national environmental groups such as Friends of the Earth, the Environmental Defence Fund, and the Sierra Club devoted some attention to genetics issues between 1976 and 1979. But the commitment by these groups was limited usually to a single staff person. Their activities were met with skepticism, and in some cases, disapproval by some board members who had ties to the biomedical community. Fourth, but to a lesser extent, litigation had drawn public attention to the issues.

Two suits were filed against HEW/NIH in 1977. A resident of Frederick, Maryland, petitioned the court to enjoin the NIH from performing a risk assessment experiment until an environmental impact statement was prepared. A second suit filed by Friends of the Earth alleged that HEW failed to comply with the requirements of the National Environmental Policy Act when it established policies for rDNA research.

Fifth, congressional interest in the safety of rDNA research intensified in the mid-to-late 1970s. House and Senate hearings provided a forum for activist scientists and environmental lobbyists. The accumulated effect of these factors resulted in heavy media coverage for both local and national events related to gene-splicing experiments.

The subsequent public debate over the specific issue of the deliberate release of genetically engineered organisms has been fostered by a similar configuration of factors, namely, scientist-critics, litigation, environmental coalitions, local opposition to experiments, and congressional interest. However, the old and new rDNA debates differ in the weighting of these factors.

For example, Jeremy Rifkin has proven particularly effective in drawing public attention to the environmental and ethical issues associated with the deliberate release of GEOs. Operating with the modest resources of his non-profit organization, Rifkin has utilized a variety of techniques, including confrontational protests, litigation, and short-term coalition building, to dramatize breaking issues in genetics and public policy.

Rifkin's philosophical opposition to genetic engineering has been expressed in his writings and media accounts that feature his work: "Genetic engineering represents the ultimate negation of nature." On genetics policy, he has had more impact on the media that any single group or individual in the United States. His notable success as a publicist and genetics critic can be explained by several factors. Rifkin can act quickly since he is not accountable to a board of overseers or a mass organization.

He is able to identify and capitalize on the weakest link of any policy process. With properly timed and targeted litigation, Rifkin has been able to strike at the jugular of the bureaucracy. He has proven time and again that a well-placed lawsuit is a magnet for media attention. Rifkin made his national debut as a genetics activist in 1977 when he and a group of supporters invaded a meeting at the National Academy of Sciences devoted to rDNA science and policy. His strategy at the NAS is best described by the 1960s term "guerrilla theater," a dramatically staged political confrontation.

Fearing major disruption of its conference, academy officials allotted Rifkin time at the start of the programme for a statement. Meanwhile, his associates, donning silk stocking face masks, stood like motionless icons, projecting a grotesque apparition of human mutants. While Rifkin took over the podium of the academy auditorium, his colleagues stretched a banner across the stage with a quote from Adolf Hitler that stated: "We will create the perfect race."

On another occasion, Rifkin organized a signature campaign of scientists and religious leaders who were asked to support a statement against genetic manipulation of the human germline. The list of supporters included individuals of vastly different ideological persuasion. To have such a politically diverse group on the same side of an issue was itself a news event.

In the fall of 1983, when the RAC was reviewing a proposal for field-testing ice minus, Rifkin brought national attention to the problem of deliberate release by filing suit against NIH on the grounds that its approval was given prior to a full study of the environmental consequences. His suit was supported by two organizations, Environmental Action and the Environmental Task Force.

To strengthen his case, Rifkin received sworn affidavits from leading ecologists who opposed the release of the ice minus strains, at least until more could be learned about the environmental consequences.

Rifkin's lawsuits have been rallying points for critical debate on genetics issues. Philosophically, he personifies the modern genetics Luddite. But on a practical level, he has made effective use of environmental law to impose a greater burden of responsibility for assessing the risks and eliciting the ethical consequences of new technologies.

Ironically, he is secretly applauded by many who, while they might differ with him philosophically, find a certain degree of relief in slowing the pace of regulatory approval. Environmental groups, skeptical about deliberate release, have been hesitant to act more aggressively because the hazards of genetic engineering are still hypothetical. For Rifkin, it was less important to prove the hazards exist than to rally support against the hubris of science and reexamine its assumptions about progress.

There is a striking comparison between the old and new genetics controversies on the role of scientific participation. In 1976, the molecular geneticists were taken to task for not calling upon infectious disease epidemiologists to evaluate the risks of transforming *E. coli* into an epidemic pathogen. Subsequently, the infectious disease community did participate in the risk evaluation. Similarly, ecologists played no role in the rDNA debate until the controversy over deliberate release erupted.

Their entry into the discussion over deliberate release came when the social and political context was ripe for their participation. Ecologists are viewed as natural allies to environmentalists; when the latter became involved in the debate over field tests, they sought advice from the former. Also, the Committee on Science and Technology of the House of Representatives made specific recommendations to NIH and USDA that they revise the memberships of their advisory committees to include individuals specifically trained in ecology and environmental sciences.

Finally, the EPA began developing its own policy on deliberate release in the early 1980s. The agency is accustomed to working with ecologists, for example, in evaluating pest control methods. The first genetically engineered organisms planned for environmental release were classified as biological control agents.

The EPA began hiring consultants to study the potential environmental problems of releasing such agents. A terrestrial ecologist from Oak Ridge National Laboratory, Frances Sharples, completed a study for the EPA in 1982 on the effects of the introduction of organisms with novel genotypes into the environment. Microbial ecologist Martin Alexander of Cornell University chaired an EPA study group on biotechnology.

In 1984, ecologists participated in an EPAsponsored workshop at the Banbury Centre at Cold Spring Harbour, New York, on the evolutionary consequences of biotechnology. Also, in 1985, with support from the EPA, the USDA, national environmental organizations, and corporations, ecologists, geneticists, public interest representatives, and federal regulators met for four days at Cornell's biological field station in Bridgeport, New York, to develop principles of containment for genetically engineered organisms tested under field conditions.

By raising the issues of biotechnology in the national media, Rifkin also helped catalyze the concerns and perspectives of ecologists on the problems of deliberate release, particularly those who had not been brought in as consultants to the major agencies. Eventually, letters from ecologists appeared in *Science,* followed by responses from geneticists. A dialogue was opened between molecular geneticists, microbiologists, and ecologists on their approaches to risk assessment.

Soon it became evident that the issues over ice minus unfolded into a larger debate about scientific culture, epistemology, and disciplinary hegemony. The next chapter looks at how the first field test of a genetically.

Engineered microorganism was evaluated by the EPA and received by the general public and citizens in communities slated for the tests.

We will see how the ice minus bacterium became an international symbol for resistance against biotechnology by some groups like Earth First and the European Greens.

To the biotechnology industry, ice minus had quite a different symbolic value as a product that could raise public confidence in a new technological frontier that promised to advance agricultural efficiency while making peace with nature.

GENETIC ALGORITHMS

Genetic programming is a branch of genetic algorithms. The main difference between genetic programming and genetic algorithms is the representation of the solution.

Genetic programming creates computer programmes in the lisp or scheme computer languages as the solution. Genetic algorithms create a string of numbers that represent the solution.

Genetic programming, uses four steps to solve problems:

1. Generate an initial population of random compositions of the functions and terminals of the problem (computer programmes).
2. Execute each programme in the population and assign it a fitness value according to how well it solves the problem.
3. Create a new population of computer programmes.
   - Copy the best existing programmes

- Create new computer programmes by mutation.
- Create new computer programmes by crossover(sexual reproduction).

4. The best computer programme that appeared in any generation, the best-so-far solution, is designated as the result of genetic programming.

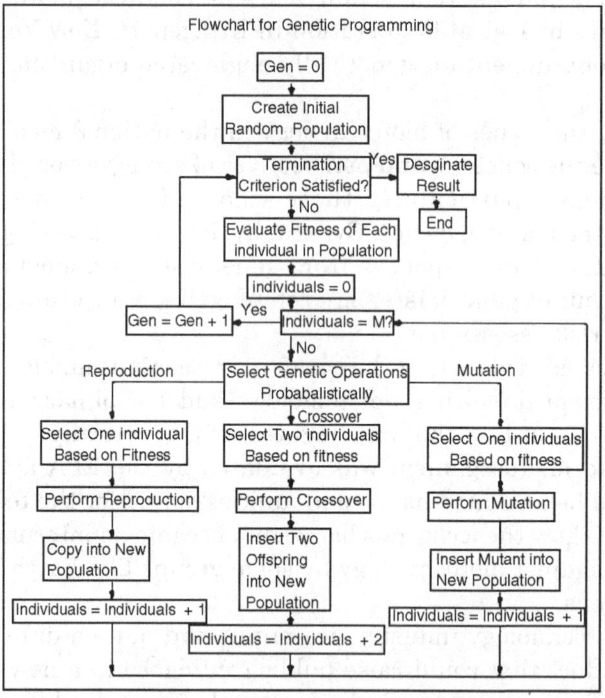

Fig. Genetic Programming Flowchart.

FITNESS FUNCTION

The most difficult and most important concept of genetic programming is the fitness function. The fitness function determines how well a programme is able to solve the problem. It varies greatly from one type of programme to the next.

For example, if one were to create a genetic programme to set the time of a clock, the fitness function would simply be the amount of time that the clock is wrong. Unfortunately, few problems have such an easy fitness function; most cases require a slight modification of the problem in order to find the fitness.

GUN FIRING PROGRAMME

A more complicated example consists of training a genetic programme to fire a gun to hit a moving target. The fitness function is the distance that the bullet is off from the target. The programme has to learn to take into account a number of variables, such as wind velocity, type of gun used, distance to the target, height of the target, velocity and acceleration of the target. This problem represents the type of problem for which genetic programmes are best. It is a simple fitness function with a large number of variables.

WATER SPRINKLER SYSTEM

Consider a programme to control the flow of water through a system of water sprinklers. The fitness function is the correct amount of water evenly distributed over the surface. Unfortunately, there is no one variable encompassing this measurement. Thus, the problem must be modified to find a numerical fitness. One possible solution is placing water-collecting measuring devices at certain intervals on the surface.

The fitness could then be the standard deviation in water level from all the measuring devices. Another possible fitness measure could be the difference between the lowest measured water level and the ideal amount of water; however, this number would not account in any way the water marks at other measuring devices, which may not be at the ideal mark.

MAZE SOLVING PROGRAMME

If one were to create a programme to find the solution to a maze, first, the programme would have to be trained with several known mazes. The ideal solution from the start to finish of the maze would be described by a path of dots. The fitness in this case would be the number of dots the programme is able to find. In order to prevent the programme from wandering around the maze too long, a time limit is implemented along with the fitness function.

FUNCTIONS AND TERMINALS

The terminal and function sets are also important components of genetic programming. The terminal and function sets are the alphabet of the programmes to be made. The terminal set consists of the variables and constants of the programmes. In the maze example, the terminal set would contain three commands: forward, right and left.

The function set consists of the functions of the programme. In the maze example the function set would contain: If "dot" then do x else do y. In the gun firing programme is the terminal set would be composed of the different variables of the problem.

Some of these variables could be the velocities and accelerations of the gun, the bullet and target. The functions are several mathematical functions, such as addition, subtraction, division, multiplication and other more complex functions.

CROSSOVER OPERATION

Two primary operations exist for modifying structures in genetic programming. The most important one is the crossover operation. In the crossover operation, two solutions are sexually combined to form two new solutions or offspring. The parents are chosen from the population by a function of the fitness of the solutions. Three methods exist for selecting the solutions for the crossover operation. The first method uses probability based on the fitness of the solution. If f $(s_i(t))$ is the fitness of the solution Si and,

$$\sum_{j=1}^{M} f\left(s_j(t)\right)$$

is the total sum of all the members of the population, then the probability that the solution Si will be copied to the next generation:

$$\frac{f\left(s_i(t)\right)}{\sum_{j=1}^{M} f\left(s_j(t)\right)}$$

Another method for selecting the solution to be copied is tournament selection. Typically the genetic programme chooses two solutions random. The solution with the higher fitness will win. This method simulates biological mating patterns in which, two members of the same sex compete to mate with a third one of a different sex. Finally, the third method is done by rank.

In rank selection, selection is based on the rank, (not the numerical value) of the fitness values of the solutions of the population. The creation of the offsprings from the crossover operation is accomplish by deleting the crossover fragment of the first parent and then inserting the crossover fragment of the second parent.

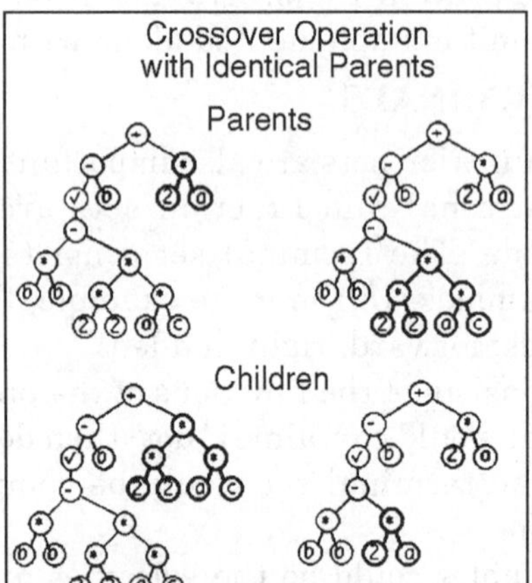

Fig. One of the Main Advantages of Genetic Programming Over Genetic Algorithms.

In genetic programming identical parents can yield different offspring, while in genetic algorithms identical parents would yield identicaloffspring. The bold selections indicate the subtrees to be swapped. An important improvement that genetic programming displays over genetic algorithms is its ability to create two new solutions from the same solution. In the Figure the same parent is used twice to create two new children.

MUTATION

Mutation is another important feature of genetic programming. Two types of mutations are possible. In the first kind a function can only replace a function or a terminal can only replace a terminal. In the second kind an entire subtree can replace another subtree. Figure explains the concept of mutation:

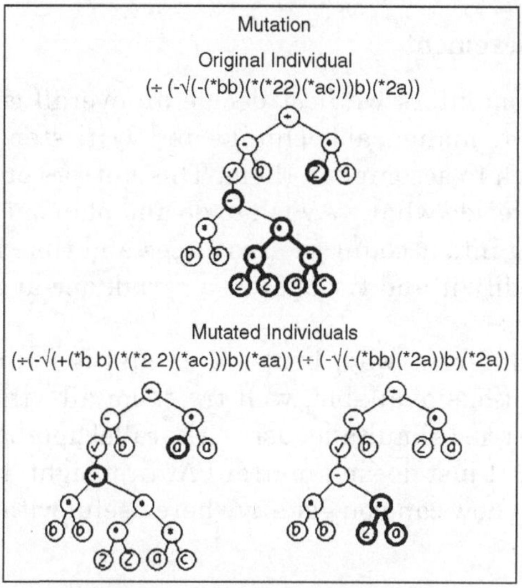

Fig. Two Different Types of Mutations.

The top parse tree is the original agent. The mutation of a single terminal for another single terminal (a). It also illustrates a mutation of a single function (–) for another single function (+). The parse tree on the bottom right illustrates a the replacement of a subtree by another subtree.

Genetic programming is much more powerful than genetic algorithms. The output of the s is a quantity, while the output of the genetic programming is a another computer programme. In essence, this is the beginning of computer programmes that programme themselves. Genetic programming works best for several types of problems.

The first type is where there is no ideal solution, (for example, a programme that drives a car). There is no one solution to driving a car. Some solutions drive safely at the expense of time, while others drive fast at a high safety risk.

Therefore, driving a car consists of making compromises of speed versus safety, as well as many other variables. In this case genetic programming will find a solution that attempts to compromise and be the most efficient solution from a large list of variables. Furthermore, genetic programming is useful in finding solutions where the variables are constantly changing. In the previous car example, the programme will find one solution for a smooth concrete highway, while it will find a totally different solution for a rough unpaved road.

GENETIC ALGORITHMS - NATURE'S WAY

In the vast eons of time the processes of evolution have created a common basis for all life on Earth - that of DNA or its simpler variant RNA. Both systems share a method of coding with that greatest of human inventions, the computer, yet the mode of operation of the two processes could hardly be more different.

Approaches to Achievement

In our use of computers we first decide an overall goal, we then break down the problem into manageable chunks and write step by step procedures (algorithms) by which to accomplish them. This mirrors our normal approach to life, in which we decide what we wish to do and plan out a way of achieving our ambition, taking into account the resources and time needed. If the plan doesn't work we modify it and try again - a serial, one at a time approach to achievement.

Nature has a different way, brute force and ignorance. If there are a hundred possible options available, well try them all - the first to get there wins. All options are tried simultaneously - a parallel approach to achievement. But where is there ? It just doesn't matter ! At first sight, to human eyes, this is a strange concept - how can you get anywhere useful without knowing where it is ?

TEACHING VERSUS EXPLORING

"Ignorance is bliss" goes the saying and we humans certainly start from a state of ignorance. Each of us can only plan for what we already know, the constraints of our limited knowledge restrict what we can imagine - we often cannot see beyond our assumptions. To overcome these limitations we must rely on being taught new ideas by someone whom we trust, or by stumbling onto new discoveries by accident - in both ways we expand our mental horizons and thus see new possibilities that were previously unthinkable. By adopting nature's method and trying every possible option we can arrive at every destination simultaneously. Thus we can enjoy every possible experience. Wow, what a lifestyle ! Is there a snag ? Of course. We consume a hundred times as many resources for every hundred possible options - and the options increase geometrically (like sand grains doubling for every square on a chessboard), the universe just isn't big enough. Note however, that the conformist social equivalent of everybody doing exactly the same thing (all making the same mistake...), uses just as many resources, yet achieves nothing better - despite the considerableduplication waste involved, our planet isn't big enough for this either !

EVALUATING OPTIONS

But nature has the solution already, tree pruning. Imagine we have two options, each of those has two more and so on, drawing those out we get a picture like a branching tree.

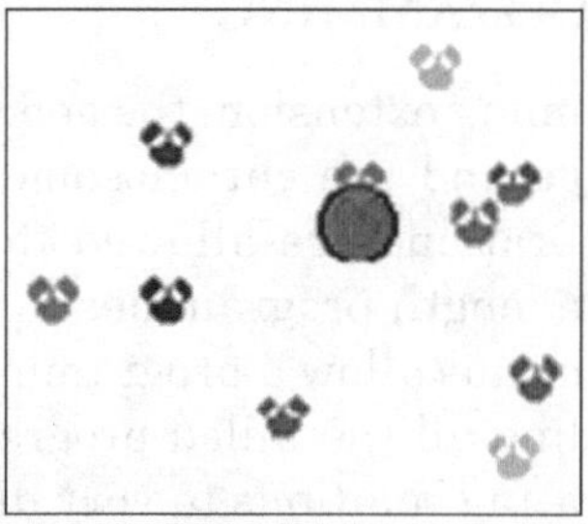

Suppose we impose a test on each option, if the test fails then that option (branch) is deleted (in this illustration we eliminate any mouse genes that fail to find the central food). Once we do this we have pruned the tree of all failed branches and those left passed all our tests (this is similar to how Chess computer programmes work). But which test does nature use ? A very simple one - survival itself. Any organisms unable to survive and reproduce will die out, their genes pruned from the available options.

Can computers operate as nature does ? Yes, this is the general field known as Evolutionary Computation, of which the Genetic Algorithm is the basic technique and can be further expanded to remove the need for an underlying programme.

THE GENETIC ALGORITHM

How does it work? Well, we create a population of individuals, each represented by a chromosome (a collection of genes or characteristics) appropriate to the problem we are trying to investigate. Genes are usually implemented by a series of computer bits, enough to cover all available alternative values. The value of each gene is initially assigned randomly, within the available parameters.

The population is then evaluated to determine how well each individual does the required task. The best members of the population (typically the top 10per cent) then become the parents for the next generation. The genes of the offspring are created by selecting two parents at random and combining part of the chromosome of each, so parent gene combinations ABC mated with DEF may become two offspring ABF and DEC. This mimics the crossover or recombination of sexual reproduction in nature and is repeated for further pairs until a full second population is created. Random changes to individual bits are then made mimicking the natural role of mutation, at some desired rate (say 1 in 100 bits changed).

The resultant population is then evaluated as before, and the process is repeated as many times as desired until the required performance level has been achieved (or no further improvement seems possible). This technique can rapidly covers the space of all possible options and converge on a solution that is beyond the ability of all but the best human programmers, in areas where no conventional solution techniques exist.

COEVOLUTIONARY PROGRAMMING

In the Genetic Programming extension, the genes are the valid statements within a computer language and the chromosome in this case is itself the programme. Additional operations are allowed to join or split programme segments, allowing variable length programmes to evolve. By GA techniques (excluding mutation) we can thus allow a programme to create itself to solve a particular problem, discarding all the failed programmes. This has proved a very powerful way of generating solutions to very difficult problems, but with the drawback that the resultant code is all but incomprehensible to humans !

There is a difficulty here. If our test is too specific, then the programme might be perfect for that one task, but fail completely for very similar tasks. How can we overcome this ? Again by copying nature. Suppose we wish to create a programme to sort a series of numbers. If we have a fixed test sequence of numbers we encounter the problem mentioned, so let us have many sets of numbers, as many as we have candidate programmes.

Programmes can then encounter different tests at each stage and only those capable of a general solution will survive. But we can do better still, let us evolve the tests also by the same process, the test survivors will the ones that defeat a programme. Using this coevolution strategy, the tests get harder as the programmes get better and this bootstrapping feedback technique gives a considerable improvement in speed and efficiency. If rather than programmes and tests we have interacting agents we enter the realm of Artificial Life.

Evolutionary Thinking

Can we use these techniques in human thought processes ? Perhaps we already do. Whenever we think of a task do we not automatically generate many possible solutions and then evaluate them (subconsciously in most cases) against our goal, discarding all those that fail our criteria ? We then refine the survivors (combining the best parts of each) and so on until we have an optimum solution. Nature has again beaten us to it...

Socially a similar process is common. Brainstorming is a popular management technique for generating diverse pseudo-random solutions to a problem, equivalent to nature's mix and match juggling of genes perhaps. The group then evaluates (tests) the options and prunes those failing some criteria, the winners then propagate to the next stage and perhaps ultimately into production. Along with top down (analysis) and bottom up (synthesis) strategies - both goal directed, we also operate a parallel evolution strategy, companies (alternative chromosomes) compete to grow, ideas (memes - cultural genes) compete for attention. Evolution is embedded in our behaviour.

ART AND INTERACTIONS

The technique of Genetic Algorithms is widely applicable, even in art. If we code genes to refer to some feature of an image or music we can evolve extraordinary organically based artforms or music. Any system of building blocks can be readily adapted to these procedures and in this way transcend automatically the imagination of its creator. By constraining the system by the desired criteria (tests) we can allow design processes to evolve as a background task - resulting in quite unexpected innovations.

Nature's way. We must however be clear that we are not fully imitating nature here. Our artificial genes (called G-Types) have well defined correspondences with function or form (called P-Types). We also have only one chromosome of very limited length in our genome, whereas natural genotypes contain many chromosomes, each with thousands of genes.In nature genes translate to body form (phenotype or morphology) via a very complex interactive process, the form is itself an emergent phenomena dependent on the interaction of genes with themselves, with the cell chemistry and with the surrounding environment.

We cannot predict this process adequately without understanding the nature of emergence. Genetic engineering, whilst highly useful, is a very dangerous process. The side effects of any changes to a such a complex and highly non-linear process as life are, to say the least, unpredictable.

PRINCIPAL DIRECTIONS OF BIOTECHNOLOGICAL DEVELOPMENT

In a recent review of agricultural technology, the Office of Technology Assessment (OTA) of the US Congress defined biotechnology to include 'any technique that uses living organisms or processes to make or modify products, to improve plants or animals or to develop micro-organisms for specific uses-it focuses upon recombinant DNA and cell fusion technologies'.

Longworth concurs with this definition, with the significant addition of tissue-culture techniques, an aspect of biotechnology which has great economic, and therefore social and political, potential impact. Despite debates about whether these or any other definitions are adequate, they do touch on the main aspects of biotechnology which deserve to be considered.

The aspect of this new biotechnology which most captures the imagination and stirs the greatest controversy is gene splicing or recombinant DNA techniques, which inspire researchers to consider the possibilities of producing reproducible animals and plants markedly different from current existing species, and referred to as transgenic species.

Already in higher animals and plants recombinant DNA has produced transgenic forms which are being commercially exploited. Their agricultural significance is, however, so far limited, and commercial application is concentrated in highly profitable pharmaceutical and horticultural markets. Genes have been introduced into several animal species which alter their protein synthesis to enable transgenic sheep to produce insulin in their milk, and rabbits to produce interferon.

A completely different application has originated in Denmark for salmon, where it has proved possible to introduce germplasm which enables the salmon's physiology to handle heavy metals, which are normally toxic, so opening up new locations for farm fisheries.

Much current research is directed to conferring disease immunity on animals, and holds out the prospect of widespread commercial application. In plants one achievement has been the transfer of genetic resistance to antibiotics in the petunia, and another has been the introduction of storage-protein genes from French bean plants into tobacco plants. As yet commercial progress with recombinant-DNA technology in plants appears limited, but extensive opportunities beckon, particularly because plant research is less restricted by the ethical and animal-welfare concerns which apply to research on transgenic animals.

The most important commercial developments based on gene splicing have so far occurred with genetically much simpler microorganisms, and it is with these that the greatest short-to mediumrun commercial potential lies. Already genetically engineered micro-organisms are producing a variety of hormones, vaccines, enzymes and other proteins. Important examples are the production of insulin, vaccines for neo-natal diarrhoea in calves and piglets, and the bovine growth hormone BST identical to that produced in cows which can stimulate a 20 per cent increase in milk yield. BST is already a source of problems for legislators in the European Community, and has provoked strong reactions from the media and milk consumers.

As far as larger animals, and cattle in particular, are concerned, it is developments in embryo transfer and in many processes for manipulating reproduction which hold out the prospect of continuing increases in yields, feed conversion efficiency and general economic efficiency. Already, apparently over 1 per cent of dairy calves in the USA are from embryo transplants, despite the high costs still associated with this procedure.

Widespread adoption of these sophisticated technologies would further distance livestock farming from its traditional rural simplicity, and from the natural mating of animals to produce offspring. It would place technical demands upon the operators which favour large-scale company farms capable of supporting a range of highly trained specialists.

Longworth identifies new techniques of tissue and cell culture as having 'the potential for enormous advances in crop improvement in the next couple of decades'. Those techniques 'can both increase the genetic diversity and greatly increase selection efficiency', and they permit innumerable plants to be reproduced asexually from single cells or small pieces of tissue. The particular technique known as 'callus culture' has been used for many years to clone highly valued horticultural plants such as orchids, and cloning of cuttings is widely practised for tree crops such as tea and palm oil as well as by millions of gardeners for garden plants.

According to Longworth 'cell culture' of single cells has unexpectedly, and so far inexplicably, resulted in plants with different properties being regenerated from the same clump of parent tissue. Sugar cane, maize and potato plants regenerated in this way have been found which are resistant to important pathogens. This application of cell culture with the capacity to generate vast numbers of seedlings rapidly has the potential to simplify greatly the hitherto labourious procedures of plant breeding and selection.

However, in terms of current and immediate commercial importance, it is through micro-organisms that biotechnology has its greatest impact. Microbial fermentation processes have been of great commercial significance for centuries, for example in bread, wine and cheese production, and there is the prospect of considerable development. Two recent examples of new processes indicate the sorts of impact that such developments can have.

In the late 1960s genetically engineered bacteria were developed which were able to digest corn starch to produce high-fructose corn syrup (HFCS) and which left as a residue corn gluten, which is now an important protein feed for livestock. HFCS has made considerable inroads in the United States and in Japan. It has largely replaced sugar as a sweetener in Coca Cola as well as many other food and drink products.

This has helped depress sugar prices to Third World growers. In the European Community steps have been taken to prevent HFCS and other new powerful sweeteners from being produced and from undermining the market for domestically grown beet sugar, which is supported by the Common Agricultural Policy.

A second important application of microbial fermentation has been the production of ethanol from sugar cane in Brazil and from maize in the USA. The commercial viability of these processes is critically dependent upon the price of oil as the main non-renewable source of fuel, and, to date, massive subsidies have been required to maintain the Brazilian and USA ethanol programmes.

Longworth states, however, that a new biotechnology, Sucrotech, is being patented, which will not only reduce the cost of producing ethanol from sugar cane, but will simultaneously produce fructose at a cost which will be competitive with HFCS and may thereby reclaim part of the sweetener market for sugar cane. That such possibilities are in prospect indicates how volatile the future might be; biotechnology has tilted the competitive balance from sugar cane to maize, causing economic pressure and even disruption to sugar-cane-dependent economies and may in future switch it back again.

In the long run the capacity to ferment fuels microbially from renewable agricultural feedstocks points to an important long-term reorientation of agriculture if non-renewable oil becomes uncompetitively expensive as a fuel for cars and feedstock for certain chemicals. It suggests that eventually there will be an increased emphasis on agricultural production of industrial feedstocks at the same time as continually increasing food output will be required to feed the expanding world population. All this will require considerable increases in agricultural productivity, to which biotechnology will increasingly contribute, but it will at the same time impose great strains on the natural environment and the structure of agriculture.

CELLULAR MECHANICS

The predominant metaphor for the cell in the age of modern biotechnology is that of a complex biochemical factory. Cells were appropriated for human use thousands of years before the industrial revolution. But not until the 1970s did scientists learn how to redesign the cell's architecture and appropriate its internal biochemical processes for production. The cell can be adapted to synthesize products that were not part of its telos or evolutionary plan. Cellular mechanics consists of those principles that define the cell's system of production.

The key discoveries that contribute to the mechanistic model of the cell span a period of nearly 40 years. In 1953 the genetic code was deciphered and the role of DNA (deoxyribonucleic acid) in the cell's synthesis of proteins was postulated and subsequently confirmed.

By 1960, messenger RNA (ribonucleic acid) was discovered. These molecules carry information from DNA to certain production sites of the cell called ribosomes where polypeptides are manufactured and fashioned into proteins. In 1967 an enzyme was isolated that causes DNA chains to join together (DNA ligase). Three years later another group of chemical enzymes were isolated called restriction enzymes (endonuclease).

These enzymes cut DNA strands at specific sites. Between 1972 and 1973 the newly discovered enzymes were used in the first experiments performed involving the introduction of foreign DNA into bacteria by cutting and splicing gene fragments. This class of experiments was named "recombinant DNA" expressing the fact that genetic information (DNA) is recombined in the test tube (*in vitro*).

The symbols, metaphors, and models that grew out of the recent discoveries in molecular biology treat the cell like a complex machine. In part, this may be explained by the desire of scientists to communicate to an audience that lacks familiarity with the arcane language of biochemistry. But it also reflects the viewpoint within science that the cell has been brought under human control. The language of the machine is the language of power, determinism, causality, matter in motion, energy, and of controlling forces.

The mechanistic materialism of modern genetics is quite distinct from the mechanistic views of physicists in the seventeenth and eighteenth centuries. The Newtonian world is characterized by material bodies in constant motion subject to external forces in a stochastic but deterministic system. The modern conception of the cell is based on a cybernetic materialism involving information transfer, energy, feedback signals, primary and secondary matter, synthesis of molecules, replication, and reproduction.

In the language of the new cybernetic materialism of molecular genetics, genes or DNA are referred to as "bits of information" or the carriers of the "code" for the production of proteins. These proteins form the "bricks" of the cell. Cells are the basic units of life; all microbes, plants, and animals are built from these tiny "chemical factories." DNA is said to "unzip."

DNA synthesizes messenger RNA (mRNA). The mRNA leaves the cell's nucleus and enters the ribosome, which begins the translation of its encoded information into chains of polypeptides that make up proteins. Ribosomes have been characterized as the cell's workrooms or its anvils where proteins are assembled.

Their action is likened to that of a sewing machine; a more appropriate metaphor might be that of a laser reader. Genomes are spoken of as being mapped, DNA as being read or sequenced. There are classes of regulator genes with well-defined functions: promoters, operators, terminators, attenuators, cap sites, retro-regulators, translational control sites, switch sites, etc.

A *Scientific American* article began with the statement: "A microorganism is a finely integrated machine that has evolved to serve its own purposes: survival and reproduction." The term "gene machines" has been introduced to describe devices that synthesize specified stretches of DNA automatically. The frequently used expression "reprogramming microorganisms" exploits the language of computers to characterize human-induced changes in the DNA "code."

THE CELL AS A FACTORY

The symbol of the cell as a machine or factory unifies several key elements of the revolution in molecular genetics. It forms a bridge between the micro and macro aspects of biotechnology. The factory metaphor can be applied to the cell, the higher organism, the farm, and eventually to the urbanization of agricultural production. Taken literally, the symbol describes the cell as a place of production. The cell's "machinery" can be appropriated to produce many useful substances, including pharmaceuticals, human proteins, and specialized chemicals, by incorporating into the cell the DNA template for the product desired. There are four requirements.

First, the foreign DNA must be incorporatable into the cell. Second, the cell must be capable of reproduction. Third, it must be capable of reproducing the DNA that has been implanted. And fourth, the inserted gene must be functional and capable of synthesizing a product.

We may draw an analogy to a factory that manufactures machine parts from metallic ore. Only one phase of the process, which involves the molds, determines whether the final product is a bolt or a drive shaft. Simply by changing the molds a new product may be fashioned from the raw materials. Similarly, the cellular machinery of the single-celled bacteria *Escherichia coli (E. coli)* can be used to synthesize human growth hormone or insulin once the gene for the appropriate protein and its regulating sequences are transplanted.

Anyone who has worked a factory assembly line knows the significance of speed-up. The line speed of the assembly process is increased thereby demanding higher worker productivity. A similar concept may be applied to bacterial production. Biotechnologists insert a gene into a bacterium that synthesizes a commercially useful product. For the process to be economically viable for largescale production the rate of biosynthesis must be sufficiently high. Industrial scientists study how to amplify or enhance the cell's products by introducing multiple copies of a gene or by controlling environmental factors. As a "factory" the cell must be brought to peak efficiency.

Bacteria are reproducible and disposable "factories" of production. Once biosynthesis is completed in a large-scale fermentation system, the spent organisms are retired from production and relegated to waste. These bacterial production units usually have a life span covering the manufacture of one product cycle. As bacterial cultures grow so rapidly, enough is set aside for the next cycle of production.

Prior to the discovery of recombinant DNA, the cell's inventory of products was the outcome of its evolutionary development. Now, scientists can retrofit the cell's production apparatus with new instructions. Some bacterial species exchange DNA naturally and therefore acquire new chemical outputs. But the cell's genetic architecture limits the kind of DNA it could incorporate under natural conditions.

A bacterium would gain no evolutionary or ecological advantage in incorporating human genes and synthesizing a human protein. The development of cellular mechanics or cell engineering is thus a critical phase in the emergence of commercial biotechnology.

GENETIC VARIATION AND NATURAL SELECTION

Recall from Inheritance that most animals, including you, are diploid, meaning that each of your traits is controlled by the interaction of at least two genes.

Mendel's work concentrated on the effect of a single gene from the mother and a single gene from the father to determine the genotype and phenotype of the offspring based on dominant and recessive genes. We now know that often multiple genes are involved in strange and intriguing ways to influence the phenotype of the offspring. More than 200 human traits are controlled by a single gene pair, such as dwarfism, cataracts, cystic fibrosis, and albinism.

Incomplete Dominance

Whenever two genes combine and the offspring's phenotype is a compromise between the effects of the two genes, then neither gene has expressed dominance over the other gene. In fact, one gene is incompletely dominant over the other gene that is inactive. Mendel's work with peas never encountered incomplete dominance because the heterozygous genotype always expressed the dominant gene in the phenotype.

F1

	b	b
B	Bb	Bb
B	Bb	Bb

Fig. Parents: BB × bb.

The active gene is apparently not strong enough to compensate for the loss of effect by the inactive allele. The classic example of *incomplete dominance* is the cross between a red snapdragon and a white snapdragon. According to Mendel, the offspring should be either red or white.

In reality they are all pink, indicating that neither allele is completely dominant but one is incompletely dominant. Therefore, a weak expression of one allele, the red one, is actually expressed in a heterozygous offspring. Incomplete dominance becomes evident in the F2 generation as detailed in the following Punnet square analysis of a fictitious trait: B = blue, b = yellow.

A cross between a blue (*BB*) parent and a yellow (*bb*) parent, *BB* × *bb*, should give all blue offspring in the F1 generation. Refer to the illustration *Parents: BB × bb*.

F2

	B	b
B	BB	Bb
b	Bb	bb

Fig. Parents: *Bb* × *Bb*.

Instead, the Bb genotype is green, a blending of the two colours. This indicates that the *B* gene is not strong enough to dominate the effect of the *b* allele completely. If two heterozygous individuals are mated, the results are even more intriguing. Refer to the illustration *Parents: Bb × Bb*.

The genotypes are *BB* (1); *Bb* (2); *bb* (1). A 1:2:1 ratio is typical of Mendelian inheritance patterns. However, the phenotypic ratio is: blue (1), green (2), yellow (1). This 1:2:1 ratio is not expected by Mendel's predictions.

Codominance

Codominance is a type of allele dominance that occurs when both alleles are active and expressed, as sometimes happens in a heterozygous individual. This phenomenon is most commonly displayed in the roan coat colour of horses and cows.

A roan colour is actually the result of a codominance by a gene for red hair and a gene for white hair. In a roan-coloured animal, there are equal numbers of both colours evenly dispersed within the coat, creating a ghostly brown colour. Another example is Erminette chickens. They are a mottled black and white colour because they have roughly equal numbers of black and white feathers. In their case, a gene for white feathers and a gene for black feathers are codominant in the heterozygous individuals.

Multiple Alleles

Although individual humans carry a maximum of two alleles that control the expression of a trait, often there are more than two different alleles in a given population that code for the same trait. When three or more alleles are available for a particular gene, the inheritance of those alleles follows a *multiple alleles* pattern.

The most common case for demonstrating multiple alleles is the inheritance of human blood types: *A*, *B*, *AB*, and *O*.

Combinations of three possible alleles give rise to the four blood types. Because you can only have two alleles for any gene, not everyone carries every type of allele available; but within a population, all of the multiple alleles are present. Sexual reproduction within that population keeps the alleles available for future generations and continues the genetic variation.

	A	B	O
A	AA	AB	AO
B	AB	BB	BO
O	AO	BO	OO

Key: AA – A; AB – AB; BB – B; OO – O
AO – A BO – B

Fig. Human population blood type combinations.

An expanded Punnet square shows the possible blood type combinations that exist in the human population. Refer to the illustration *Human population blood type combinations*. Remember that although there are three possible alleles listed on each axis, a human can only possess two of these. However, other individuals in the human population may carry a different set of two alleles, so that all three types are present and active in the human population.

Linked Genes

Each chromosome normally carries thousands of genes. When certain neighbouring genes on the same chromosome are inherited together, they are considered. Linked genes are an anomaly in Mendel's theory of independent assortment because they tend to remain together as a segment and therefore function as a cluster of genes rather than individual genes. In humans, the autosomal genes for red hair, freckles, and a fair complexion are linked genes.

Sex-Linked, or *X*-Linked Genes

The linkage in *sex-linked traits* is completely different from linked genes. A sex-linked trait is one whose allele is located on one of the sex chromosomes, usually the *X* for autosomal traits. Most sex-linked genes are recessive and are only expressed in homozygous females and males.

Because males normally inherit one *X* chromosome from the mother only, the recessive traits found on the *X* chromosome are passed from mother to son. It is more difficult to pass an *X*-linked trait to a daughter because she inherits an *X* from both parents, so both parents would have to possess and pass on the recessive gene.

The recessive gene on the *X* chromosome is expressed in males because it is the only gene present. In females, it can be hidden by the presence of a dominant gene on the other *X* chromosome. In humans, sex-linked traits include colour blindness, albinisms, and haemophilia. Refer to the illustration *Colour blindness pedigree*. A heterozygous *"carrier"* mates with a *"carrier"* female (X^c = gene for colour blindness). A carrier is an individual who is heterozygous for a condition but does not display the trait. Therefore, a carrier for colour blindness is not colour blind but can pass on the gene for colour blindness.

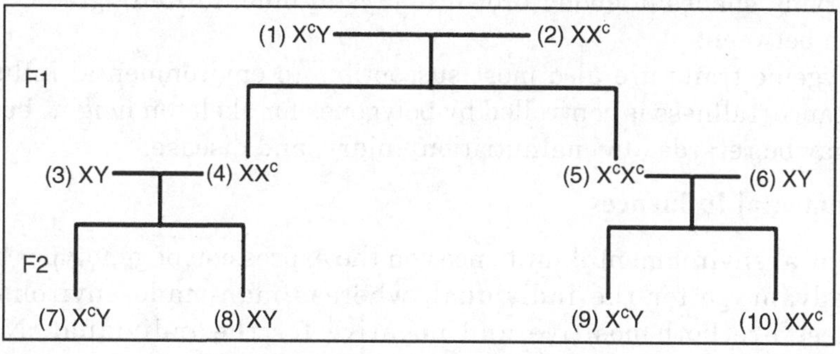

Fig. Colour Blindness Pedigree.

In the F1 generation, one colour blind female and one carrier female each mate with a non-colour blind male. In the F2 generation, individuals 7 and 8 are brothers, one is colour blind, whereas other is not. The colour blind male inherited the colour blind gene from the heterozygous carrier mother; the non-colour blind son inherited the normal *X* chromosome that did not contain the gene for colour blindness. Likewise, individual 9 could only receive a gene for colour blindness from his mother, number 5. The sister, number 10, is a heterozygous carrier.

Sex-Influenced Genes

Certain genes are activated by hormones secreted by the sex organs, so they display different phenotypes when found in males than they do when found in females, even though their genotypes are identical. The standard sex-influenced study in humans is male pattern baldness, which is influenced by testosterone and can affect both males and females. However, it is dominant in males and recessive in females!

In the homozygous state, both males and females become bald; in the heterozygous genotype, females show no signs of baldness, but males become bald. Interestingly, sex-influenced genes are generally not found on the sex chromosomes, so both sexes can have the same genotype and different phenotypes!

Polygenic inheritance is the opposite of *pleiotropic* inheritance, in which a single gene affects several characteristics.

POLYGENIC INHERITANCE

So far, we have limited gene expression to those traits expressed by the action of one gene. In most primates, including you, traits are governed by the interaction of multiple genes. The genes of *polygenic traits* may be interspersed on the same chromosome or on completely different chromosomes. The additive effects of numerous genes on a single phenotype create a continuum of possible outcomes.

A typical example is eye colour in humans, which under normal genotypes is in varying shades of ebony brown to crystal blue, to Kelly green, and all points in between.

Polygenic traits are also most susceptible to environmental influences. For instance, tallness is controlled by polygenes for skeleton height, but their effect may be retarded by malnutrition, injury, and disease.

Environmental Influences

Natural environmental influences on the expression of genotypes tend to be an advantage for the individual, whereas man-made environmental influences are both positive and negative for the individual. Natural environmental influences include the phenomenon of colour change in the Arctic fox from red-brown in the summer months to pure white during the winter season for better camouflage.

The genes that produce the red-brown summer pigment are blocked by cold temperatures, causing the hair to grow with no colour (therefore, white). Another colourful example is the interaction between the colour of the hydrangea flower, which is blue in acidic soils and pink in alkaline soils. A recent study also linked improved diet in infants and adolescents to a taller average height in the United States and Europe with the opposite effect in famine-stricken populations. The genetic complement of an individual is inherited; however, the environmental effect on these genes may alter their application and expression.

VARIATION

Variation means the differences in characteristics (phenotype) within a species.

There are many causes of variation as this chart shows:

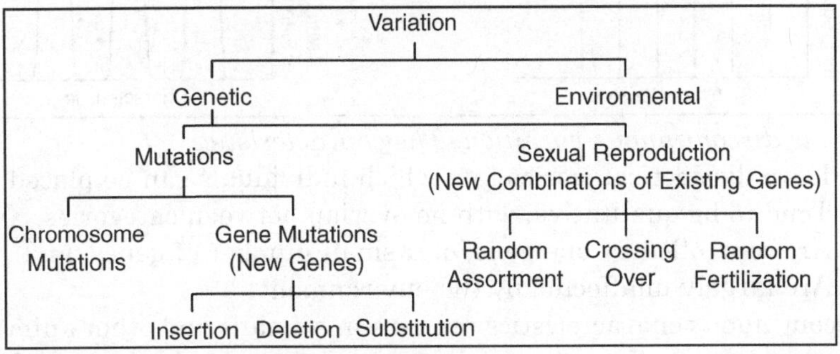

Variation consists of differences between species as well as differences within the same species. Each individual is influenced by the environment, so this is another source of variation.

Genotype + environment = phenotype

The alleles that are expressed in the phenotype can only perform their function efficiently if they have a supply of suitable substances and have appropriate conditions. Any study of variation inevitably involves the collection of large quantities of data.

When collecting the data there is a need for *random sampling*. This involves choosing a sample "*at random*" from a population. This means that every member of the population has the same chance of being chosen, and that the choices are independent of one another.

In choosing a random sample some things are very important:

- The selection criteria must not correspond to what you are studying, *i.e.*, not introduce any systematic bias.
- You must try to maintain independence of the sample points, and
- You must be careful in making inference. The population might not be what you wish it were.

Variation in a population can be studied by measuring the characteristic (height, eye colour, seed shape, or whatever) in a large number of different individuals and by then plotting a *frequency histogram*. This graph has the values of the characteristic on the X-axis (grouped into bins if necessary) and the number of individuals showing that characteristic on the Y-axis. These histograms show that there are two major types of variation: discontinuous and continuous.

DISCONTINUOUS VARIATION

Sometimes the characteristic has just a few discrete categories (like blood group). The frequency histogram has separate bars (or sometimes peaks).

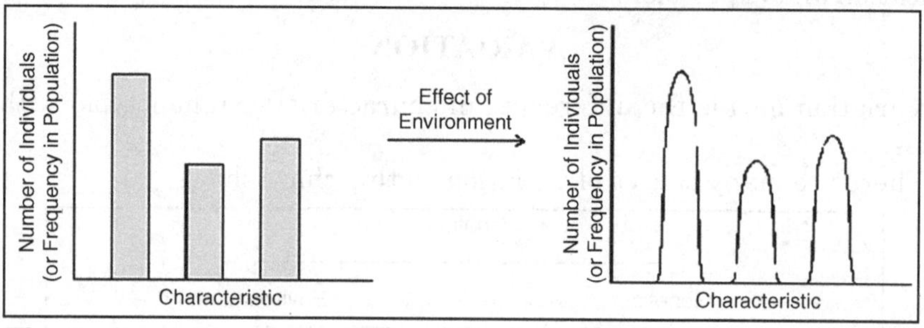

This is discontinuous variation. The characteristics:

- Have distinct categories into which individuals can be placed,
- Tend to be qualitative, with no overlap between categories,
- Are controlled by one gene, or a small number of genes, and
- Are largely unaffected by the environment.

Discontinuous characteristics are rare in humans and other animals, but are more common in plants. Some examples are human blood group, detached ear lobes, flower colour, seed colour, etc.

These characteristics are very useful for geneticists because they give clear-cut results.

CONTINUOUS VARIATION

Sometimes, the character has a continuous range of values (like height). The frequency histogram is a smooth curve (usually the bell-shaped normal distribution curve).

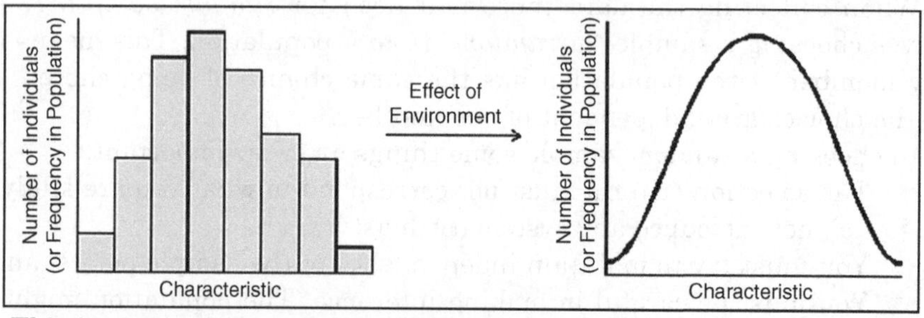

This is continuous variation. The characteristics:

- Have no distinct categories into which individuals can be placed,
- Tend to be quantitative, with overlaps between categories,
- Are controlled by a large number of genes (polygenic), and
- Are significantly affected by the environment.

Continuous characteristics are very common in humans and other animals. Some examples are height, hair colour, heart rate, muscle efficiency, intelligence, growth rate, rate of photosynthesis, etc. Characteristics that show continuous variation are controlled not by one, but by the combined effect of a number of genes, called *polygenes*. Therefore, any character that results from the interaction of many genes is called a *polygenic character*. The random assortment of genes during prophase I of meiosis ensures that individuals possess a range of genes from any polygenic complex.

Sometimes, you can see the effect of both variations. For example, the histogram of height of humans can be bimodal (*i.e.* it has got two peaks). This is because the two sexes (a discontinuous characteristic) each have their own normal distribution of height (a continuous characteristic).

Standard Deviation and Standard Error

Standard deviation is a measure of the spread of results at either side of the mean (average).

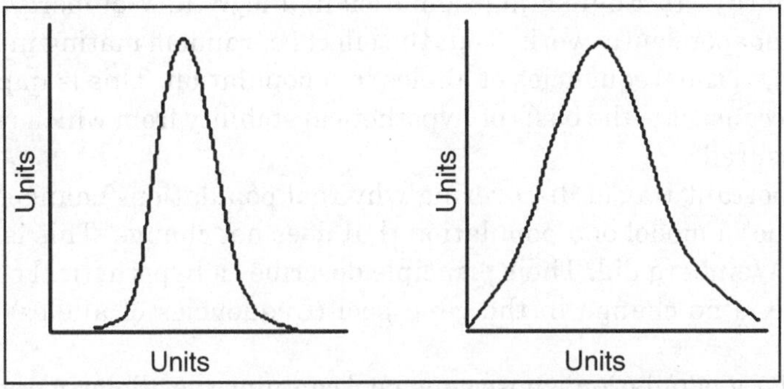

- These sets of data have the same mean (average).
- The data shows a *normal distribution* about the mean value – there is a bell-shaped and even distribution of values above and below the mean.
- The diagram on the left shows a smaller standard deviation, indicating there is less variation between individuals for the character mentioned.
- The diagram on the right shows a greater standard deviation, indicating there is greater variation.

The standard error is the standard deviation of the mean:

- If individuals in a number of samples are measured, then each sample will have its own mean.
- These means will usually be slightly different from each other – reflecting chance differences in samples – giving a range of values for sample means.
- Standard error is a measure of how much the value of a sample mean is likely to vary.
- The greater the standard error, the greater the variation of the mean.

Variation, Gene Pool, and Populations

It is at the population level that evolution occurs. A population is a group of individuals of the same species in a given area whose members can interbreed. Because the individuals of a population can interbreed, they share a common group of genes known as the *gene pool.* Each gene pool contains all the alleles for all the traits of all the population.

For evolution to occur in real populations, some of the gene frequencies must change with time. The *gene frequency* of an allele is the number of times an allele for a particular trait occurs compared to the total number of alleles for that trait.

$$\text{Gene Frequency} = \frac{\text{the number of specifice types of allele}}{\text{the total number of alles in the gene pool}}$$

Hardy–Weinberg Equation

G.H. Hardy, an English mathematician, and W.R. Weinberg, a German physician, independently worked out the effects of random mating in successive generations on the frequencies of alleles in a population. This is important for biologists because it is the basis of hypothetical stability from which real change can be measured.

An important way of discovering why real populations change with time is to construct a model of a population that does not change. This is just what Hardy and Weinberg did. Their principle describes a hypothetical situation in which there is no change in the gene pool (frequencies of alleles), hence no evolution.

Consider a population whose gene pool contains the alleles *A* and *a*. Hardy and Weinberg assigned the letter p to the frequency of the dominant allele *A* and the letter q to the frequency of the recessive allele *a*. Since the sum of all the alleles must equal 100 per cent, then $p + q = 1$. They then reasoned that all the random possible combinations of the members of a population would equal $(p + q)^2$ *or* $p^2 + 2pq + q^2$.

The Hardy-Weinberg equation:

$$p^2 + 2pq + q^2 = 1$$

The frequencies of A and a will remain unchanged generation after generation if the following conditions are met:

1. *Large population:* The population must be large to minimize random sampling errors.
2. *Random mating:* There is no mating preference. For example, an *AA* male does not prefer an *aa* female.
3. *No mutation:* The alleles must not change.
4. *No migration:* Exchange of genes between the population and another population must not occur.
5. *No natural selection:* Natural selection must not favour any particular individual.

Let us look at an analogy to help demonstrate Hardy-Weinberg Equilibrium. Imagine a "*swimming*" pool of genes.

Find: Frequencies of *A* and *a*. and the genotypic frequencies of *AA*, *Aa*, and *aa*.

Solution:

f(*A*) = 12/30 = 0.4 = 40%

f(*a*) = 18/30 = 0.6 = 60%

then, $p + q = 0.4 + 0.6 = 1$

and $p^2 + 2pq + q^2 = AA + Aa + aa$

=.24 +.48 +.30 = 1

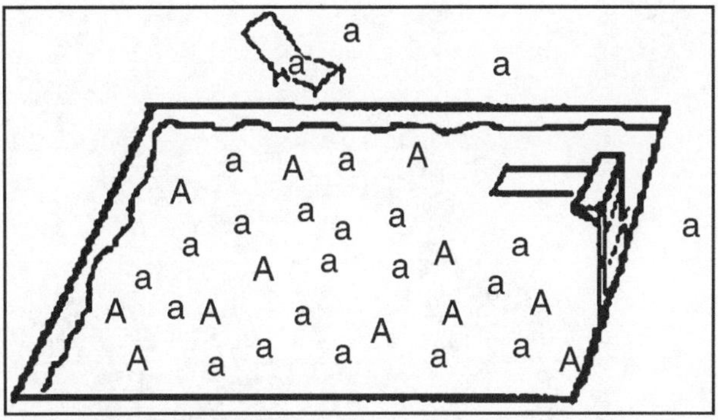

As long as the conditions of Hardy-Weinberg are met, the population can increase in size and the gene frequencies of *A* and *a* will remain the same. Thus, the gene pool does not change.

Now, suppose more "*swimmers*" dive. What will the gene and genotypic frequencies be?

Solution:

f(*A*) = 12/34 =.35 = 35 %

f(*a*) = 22/34 =.65 = 65%

f(*AA*) =.*12*, f(*Aa*) =.23 and f (*aa*)

=.42

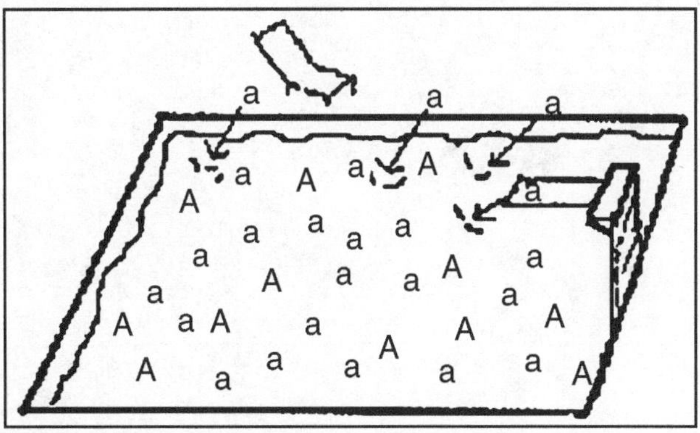

The results show that Hardy-Weinberg equilibrium was not maintained. The migration of swimmers (genes) into the pool (population) resulted in a change in the population's gene frequencies.

If the migrations were to stop and the other agents of evolution (*i.e.*, mutation, natural selection, and non-random mating) did not occur, then the population would maintain the new gene frequencies generation after generation. It is important to note that a fifth factor affecting gene frequencies is *population size*. The larger a population is, the number of changes that occur by chance alone becomes insignificant. In the analogy above, a small population was deliberately used to simplify the explanation.

8

Transformation in Genetic Engineering of Plants

INTRODUCTION

Genetic engineering of plants is much easier than that of animals. There are several reasons for this:

- There is a natural transformation system for plants (the bacterium *Agrobacterium tumefaciens*),
- Plant tissue can redifferentiate (a transformed piece of leaf may be regenerated to a whole plant), and
- Plant transformation and regeneration are relatively easy for a variety of plants.

The soil bacterium *Agrobacterium tumefaciens* ("tumefaciens" meaning tumor-making) can infect wounded plant tissue, transferring a large plasmid, the Ti plasmid, to the plant cell. Part of the Ti (tumor-inducing) plasmid apparently randomly integrates into the chromosome of the plant.

The integrated part of the plasmid contains genes for the synthesis of (1) food for the bacterium, and (2) plant hormones. Genes from the Ti plasmid that are integrated in the plant chromosome are expressed at high levels in the plant. Overproduction of the plant hormones leads to continuous growth of the transformed cells, causing plant tumors. Rapid, cancerous growth of the transformed plant tissue obviously is advantageous to the bacterium: more food gets produced. The Ti plasmid has been genetically modified ("disarmed") by deleting the genes involved in the production of bacterial food and of plant hormones, and inserting a gene that can be used as a selectable marker. Selectable marker genes generally are coding for proteins involved in breakdown of antibiotics, such as kanamycin. Any gene of interest can be inserted into the Ti plasmid as well. In principle, one can thus transform any plant tissue, and select transformants by screening for antibiotic resistance. However, unfortunately, there are some complications:

- It has proven difficult to transform some monocots (grasses, etc.) by *Agrobacterium*, and

- Regeneration of plants from tissue culture or leaf discs is not always possible.

A number of genetically engineered plant varieties have been developed. Traits that have been introduced by transformation include herbicide resistance, increased virus tolerance, or decreased sensitivity to insect or pathogen attack. Traditionally, most of such genetically engineered plants were tobacco, petunia, or similar species with a relatively limited agricultural application. However, during the past decade it now has become possible to transform major staples such as corn and rice and to regenerate them to a fertile plant. Increasingly, the transformation procedures used do not depend on *Agrobacterium tumefaciens*.

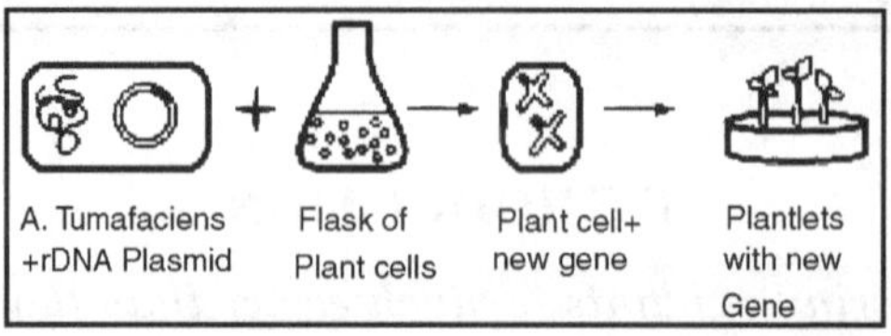

Fig. Transformation Using Ti Plasmid

Instead, DNA can be delivered into the cells by small, μm-sized tungsten or gold bullets coated with the DNA. The bullets are fired from a device that works similar to a shotgun. The modernized device uses a sudden change in pressure of He gas to propel the particles, but the principle of "shooting" the DNA into the cell remains the same. This DNA-delivery device is nicknamed "gene gun", and has been shown to work for DNA delivery into chloroplasts as well.

Over the last several years, use of the "gene gun" has become a very common method to transform plants, and has been shown to be applicable to virtually all species investigated. For example, transformation of rice by this method is now routine. This is a very important development as rice is the most important crop in the world in terms of the number of people critically dependent on it for a major part of their diet. Another method to get foreign genes into cereals is by electroporation: a jolt of electricity is used to puncture self-repairing holes in protoplasts (*i.e.*, the cell without the cell wall), and DNA can get in through these holes. However, it is often very difficult to regenerate fertile plants from protoplasts of cereals. Nonetheless, significant advances in overcoming these practical difficulties have been made over the years.

Now even transgenic trees have been created: for example, the gene for a coat protein of the plum pox virus has been introduced into apricot. The plum pox virus leads to the feared Sharka disease, for which there is no cure. The resulting transgenic tree shows a markedly decreased sensitivity to this virus. The reason why continuous exposure of the tree to the viral coat protein leads to tolerance against viral infection is not yet understood, however. Thus, now there are a number of different techniques to introduce foreign genes into plants. Essentially all major crop plants can be (and have been or are being) genetically engineered, the procedures are now routine and the frequency of success is very high. Even though genetically engineered crops are more costly than the usual ones, US farmers have rather readily accepted them provided that tangible benefits can be demonstrated. However, it is questionable whether the farmer in poorer countries can come up with the funds to "try out" and use the new crops. Another issue in this respect is how genetically engineered crops are perceived by the consumer. Even though in the US there is little resistance to such crops as long as the products can be shown to be safe and advantageous, in other countries genetically modified foods are received poorly by the consumer.

It is unlikely that there is a rationally sound basis for this rather hostile reaction of the consumer, as most of the crops are the results of human manipulation (such as centuries of breeding) and may have been treated with harmful herbicides and pesticides. Time and education will need to be invested to provide consumers and consumer advocates with a balanced opinion on the acceptability of the origin of their foods.

On the other hand, as so many plants (soybean, corn, etc.) are genetically modified and the nature of the genetic modification is not necessarily easy to explain, it may be simpler to label those foods that are guaranteed free of "genetically modified organisms" or their products. However, keep in mind that traditional breeding has genetically modified essentially all agricultural products, so it may be difficult to define what is actually free of genetically modified organisms.

TRANSGENIC CROPS

The varieties created through the use of a biotechnological approach commonly known as recombinant DNA, genetic engineering (GE), transgenic modification, or genetic modification (GM). The products of genetic engineering are often called genetically modified organisms, or GMOs. All these terms refer to methods of recombinant DNA technology by which biologists splice genes from one or more species into the DNA of crop plants to transfer chosen genetic traits.

Genes are segments of DNA that controlled the information that determines the structure and function of a living organism. Genetic engineers are molecular biologists who manipulate this information, typically by taking genes from one species—an animal, plant, bacterium, or virus—and inserting them into another species, such as an agricultural crop. With the advent of genetic engineering of plants around 1983, it appeared that this new biotechnology would benefit and even revolutionize agriculture.

The transfer of desirable genetic traits across species barriers has shown promise for solving problems in the management of agricultural crops. Potential benefits include reduced toxic pesticide use, improved weed control resulting in less tillage and soil erosion, and water conservation.

GENE FLOW TO NEIGHBOURING CROPS AND TO RELATED WILD SPECIES

Ecological scientists have little doubt that gene flow from transgenic fields into conventional crops and related wild plants will occur. Gene flow from transgenic to conventional crops is of concern to farmers because of its potential to cause herbicide resistance in related conventional crops.

For example, in western Canada, three different herbicide-resistant canola varieties have cross-pollinated to create canola plants that are resistant to all three types of herbicide. Through what is called gene-stacking, this new triple resistance has turned volunteer canola into a significant weed problem.

Gene flow from transgenic crops to wild relatives creates a potential for wild plants or weeds to acquire traits that improve their fitness, turning them into "super weeds." For example, if jointed goatgrass—a weedy relative of wheat—acquires the herbicide-tolerant trait of Roundup Ready wheat, it will thrive in crop fields unless applications of other herbicides are made. There is already evidence of such outcrossing from herbicide-resistant wheat to jointed goatgrass. Frank Young and his colleagues at Washington State University found that imidazolone-resistant wheat (not a transgenic variety) outcrossed to goatgrass in one season. Other traits that wild plants could acquire from transgenic plants that would increase their weediness are insect and virus resistance.

Because of their experience with classically bred plants, few scientists doubt that genes will move from crops into the wild: seven of the world's thirteen most important crop weeds have been made weedier by genes acquired from classically bred crops. Because gene flow has the potential to affect farmers' crop and pest management, crop marketability, and liability, more research needs to be done to determine the conditions under which gene flow from transgenic plants is likely to be significant.

Pesticide Resistance in Insect Pests

Bt has been widely used as a microbial spray because it is toxic only to caterpillars. In fact, it is a pest management tool that organic farmers depend on—one of the few insecticides acceptable under organic rules.

Unlike the commercial insecticide spray, the Bt engineered into crop plants is reproduced in all, or nearly all, the cells of every plant, not just applied on the plant surface for a temporary toxic effect. As a result, the possibility that transgenic Bt crops will accelerate insect pests' development of resistance to Bt is a serious concern. Pest resistance to Bt would remove this valuable and environmentally benign tool from farmers' and forest managers' pest control toolbox. Antibiotic resistance The use of antibiotic-resistant marker genes for the delivery of a gene package into a recipient plant carries the danger of spreading antibiotic-resistant bacteria. The likely result will be human and animal health diseases resistant to treatment with available antibiotics.

Research is needed on antibiotic resistance management in transgenic crops. Already the European Commission's new rules governing transgenic crops stipulate phasing out antibiotic-resistant marker genes by the end of 2004.

Effects on Beneficial Organisms

Evidence is increasing that transgenic crops—either directly or through practices linked to their production—are detrimental to beneficial organisms. New studies are finding that Bt crops exude Bt in concentrations high enough to be toxic to some beneficial soil organisms. In addition, a study by University of Arkansas scientists has shown that "root development, nodulation and nitrogen fixation are impaired" in some varieties of Roundup Ready soybeans.

The reason is that the beneficial rhizobium responsible for nitrogen fixation in soybeans is sensitive to Roundup. It also appears that disruption of beneficial soil organisms can interfere with plant uptake of phosphorus, an essential plant nutrient. Insecticidal crops in two ways can affect beneficial insects that prey on insect pests. First, the Bt in transgenic insecticidal crops has been shown in some laboratory studies to be toxic to ladybird beetles, lacewings, and monarch butterflies. The extent to which these beneficials are affected in the field is a matter of further study. Second, because the insecticidal properties of Bt crops function even in the absence of an economic threshold of pests, Bt crops potentially can reduce pest populations to the point that predator species are negatively affected.

Reduced Crop Genetic Diversity

As fewer and larger firms dominate the rapidly merging seed and biotechnology market, transgenic crops may continue the trend towards simplification of cropping systems by reducing the number and type of crops planted. In addition, seed saving, which promotes genetic diversity, is restricted for transgenic crops.

Food Safety

Because food safety is predominantly a concern of consumers, However, because food safety issues do impact marketing and international trade, here is a short list of consumer safety concerns about transgenic food:

- Possibility of toxins in food
- Possibility of new pathogens
- Reduced nutritional value
- Introduction of human allergens
- Transfer of antibiotic resistance to humans
- Unexpected immune-system and genetic effects from the introduction of novel compounds

Farm Management Issues

The most widely planted transgenic crops on the market today can simplify short-term pest management for farmers and ranchers. In the case of herbicide-tolerant crops, ideally farmers can use a single broad-spectrum herbicide for all their crop weeds, though they may need more than one application in a season. By planting insecticidal crops, farmers can eliminate the need to apply pesticides for caterpillar pests like the European corn borer or the cotton bollworm, though they still have to contend with other crop pests.

While these crops offer handy pest control features, they may complicate other areas of farm management. Farmers who are growing both transgenic and conventional varieties of the same crop will need to segregate the two during all production, harvesting, storage, and transportation phases if they are selling into differentiated markets or plan to save their own seed from their conventional crops.

To minimize the risk of gene flow from transgenic to adjacent conventional crop fields, federal regulations require buffer strips of conventional varieties around transgenic fields. Different transgenic crops require different buffer widths.

Because the buffer strips must be managed conventionally, producers have to be willing to maintain two different farming systems on their transgenic fields. Crops harvested from the buffer strips must be handled and marketed as though they are transgenic. Planted refuges—where pest species can live outside fields of insecticidal and herbicide-tolerant transgenic crops—are also required to slow the development of weed and insect pest resistance to Bt and broad-spectrum herbicides.

These refuges allow some individuals in the pest population to survive and carry on the traits of pesticide susceptibility. Requirements governing the size of refuges differ according to the type of transgenic crop grown.

Farmers growing herbicide-tolerant crops need to be aware that volunteer crop plants the following year will be herbicide-resistant. This herbicide resistance makes no-till or direct-seed systems difficult because volunteers can't be controlled with the same herbicide used on the rest of the crop. In a no-till system that relies on the same broad-spectrum herbicide that the volunteer plants are resistant to, these plants will contaminate the harvest of a following conventional variety of the same crop, a situation farmers need to avoid for two reasons.

First, the contamination means a following conventional crop will have to be sold on the transgenic market. This leads to the second reason. If farmers grow and market a transgenic crop for which they do not have a technology agreement and did not pay royalty fees, they face possible prosecution by the company that owns the transgenic variety.

Many farmers have been charged with "theft" of a company's patented seed as a result of contamination in the field. Farmers growing insecticidal crops need to recognize that insect pressure is difficult to predict and may not warrant the planting of an insecticidal variety every year. In a year when pest pressure is low, the transgenic seed becomes expensive insurance against the threat of insect damage.

Farmers growing transgenic crops need to communicate with their neighbours to avoid contaminating their fields and to ensure that buffers are adequate. In Maine, farmers growing transgenic crops are now required by law to be listed with the state agriculture department, to help identify possible sources of cross-contamination when it occurs. The new law, L.D. 1266, also "requires manufacturers or seed dealers of genetically engineered plants, plant parts or seeds to provide written instructions to all growers on how to plant, grow, and harvest the crops to minimize potential cross-contamination of non-genetically engineered crops or wild plant populations". Farm management issues common to all transgenic crops include yield, cost, price, profitability, management flexibility, sustainability, market acceptance, and liability.

CROP YIELD, COSTS, AND PROFITABILITY

Some farmers will get higher yields with a particular transgenic crop variety than with their conventional varieties, and some will get lower yields. The same is true for costs. Some producers will see their overall profitability rise, and some will see it drop. If the results of research studies so far on yield, chemical use, costs, crop prices, and profitability of these crops seem contradictory, it is because farmers and farms are not all alike.

Neither are the conventional varieties used for comparison. Some yield, cost, and profitability trends do appear to be emerging from the growing body of research data for transgenic crops, however. As noted in the Wallace Centre report, Roundup Ready soybeans were designed simply to resist a particular chemical herbicide, not to increase yields. In contrast, Bt corn and cotton, by resisting insect pests, may result in higher yields from reduced pest pressure.

Herbicide-tolerant soybeans appear to suffer what's referred to as a "yield drag." Again, in some areas and on some farms this tendency of Roundup Ready soybean varieties to yield less than their comparable, conventional counterparts varies, but overall, they appear to average yields that are five to ten per cent lower per acre.. Drought conditions worsen the effects. The bacterium that facilitates nodulation and nitrogen fixation in the root zone apparently is sensitive to both Roundup and drought.

Bt cotton, especially, outpaces yields of conventional cotton by as much as 9 to 26 per cent in some cases, though not at all in others. Yield increases for Bt corn have not been as dramatic. Time will tell whether farmers can expect yield increases or decreases in the long run with these and other transgenic crop varieties.

One of the promises of biotechnology is that it will reduce pesticide use and thereby provide environmental benefits and reduce farmers' costs. The herbicide-tolerant and insecticidal varieties are designed specifically to meet these goals. Increased future pesticide use resulting from the buildup of resistance to heavily used herbicides is a long-term concern. Pesticide use depends on the crop and its specific traits; weather; severity of pest infestations; farm management; geographic location of the farm; and other variables. As a result, conclusions drawn by various studies analysing pesticide use on transgenic crops remain controversial. According to the Wallace Centre report, in a review of the data available up through 2000, crops engineered to contain Bt appear to have decreased the overall use of insecticides slightly, while the use of herbicide-resistant crops has resulted in variable changes in overall herbicide use, with increases in use of some herbicides in some places and decreases in others

Farmers need to consider all the factors that determine profitability. No single factor can tell the whole story. Transgenic crop seeds tend to be more costly, and farmers have the added expense of a substantial per-acre fee charged by the owners of transgenic varieties. These costs have to be considered along with input cost changes—whether herbicide or insecticide use and costs go down, go up, or stay the same. Market price is another factor: Prices for some transgenic crops in some markets are lower than prices for comparable conventional crops, though rarely are they higher. Farmers need to watch the markets. Some buyers will pay a premium for a non-transgenic product, though as transgenic seeds find their way into conventional transportation, storage, and processing streams, these premiums may disappear along with confidence that "GMO-free" products are in fact truly free of engineered genes.

Buyer acceptance is a significant marketing issue for farmers raising transgenic crops. Farmers need to know before they plant what their particular markets will or won't accept. Since most grain handlers cannot effectively segregate transgenic from non-transgenic crops in the same facility, many companies are channeling transgenic crops into particular warehouses. Farmers need to know which ones and how far away those are.

While the European Union officially lifted its two-year moratorium on the introduction of new transgenic crops in February 2001, during the debate over labeling and traceability regulations the moratorium remains in effect. Under the proposed new EU requirements, "all foods and animal feed derived from GMOs have to be labeled and, in the case of processed goods, records have to be kept throughout the production chain allowing the GMO to be traced back to the farm".

Liability

The need to separate transgenic crops from conventional and organic ones opens farmers to liability for their product at every step from seed to table. Effective systems for segregation do not exist at present, and will be costly to develop and put into place. Farmers may well end up bearing the added costs of crop segregation, traceability, and labeling. In the meantime, farmers who grow transgenic varieties—and, ironically, those who do not—are liable for transgenic seeds ending up where they aren't wanted: in their own non-transgenic crop fields, in neighbours' fields, in truckloads of grain arriving at the elevator, in processed food products on retail shelves, and in ships headed overseas.

Farmers who choose not to grow transgenic varieties risk finding transgenic plants in their fields anyway, as a result of cross-pollination via wind, insects, and birds, which may bring pollen from transgenic crops planted miles away. Besides pollen, sources of contamination include contaminated seed and seed brought in by passing trucks or wildlife. Those farmers whose conventional or organic crops are contaminated, regardless of the route, risk lawsuits filed against them by the companies that own the proprietary rights to seed the farmer didn't buy. Likewise, farmers who grow transgenic crops risk being sued by neighbours and buyers whose non-transgenic crops become contaminated.

Because contamination by transgenic material has become so prevalent in such a short time, all farmers in areas of transgenic crop production are at risk. Insurance, the most common recourse for minimizing potential losses because of liability, is not available to the nation's farmers for this risk because insurance companies do not have enough information to gauge the potential losses. A full risk assessment and legal clarification of the distribution of liability among farmers, seed companies, grain handlers, processors, and retailers is needed before farmers can rest assured that transgenic crops won't result in lawsuits against them.

For farmers who have adopted transgenic technology—and for those who have not—in the words of Brian Leahy, executive director of California Certified Organic Farmers, "This technology does not respect property rights". In essence, farmers who grow transgenic crops on some of their fields, and farmers who grow none, risk bearing tremendous liability. This situation won't change until farmers either gain legal protection or until they all grow transgenic crops on all their fields.

Influences on Public Research

While transgenic crop varieties are generally the property of private corporations, those corporations often contract with public-sector agricultural research institutions for some of their development work. In fact, private investment in agricultural research, including germplasm development, has surpassed public investment in recent years. With this shift in funding sources, the following questions become important: Is the private sector unduly influencing the public research agenda? Are corporations directing public research in less socially valuable directions while research on, for instance, sustainable agriculture goes wanting? Are the outcomes of corporate-funded transgenic research and development by our public institutions equitable across the food and agricultural sectors? Is equity even a consideration of our public institutions when they accept this work?

Because intellectual property rights (patents) apply to living organisms, making them private property, the free flow of scientific information that has historically characterized public agricultural research is being inhibited. What are the implications for the future of agriculture and society of the secrecy that now surrounds so much of what was formerly shared public knowledge? For a brief history of intellectual property rights as they apply to living organisms. These and other questions need to be addressed by citizens and their public institutions. These issues are of particular concern to farmers and consumers who would benefit from research into alternative technologies that are less costly (in every way), less risky, and more equitable.

Equity requires that the economic benefits and risks of transgenic technology be fairly distributed among technology providers, farmers, merchants, and consumers. For long-term sustainability, farmers need research that focuses on farms as systems, with internal elements whose relationships can be adjusted to achieve farm management goals.

In contrast, transgenic crop research so far has focused on products that complement toxic chemical approaches to control of individual pest species. These are products that can be commercialized by large agribusiness or agri-chemical interests and that farmers must purchase every year. This research orientation only perpetuates the cost-price squeeze that continues to drive so many out of farming.

Industry Concentration and Farmers' Right to Save Seed

The broadening of intellectual property rights in 1980 to cover living organisms, including genes, has resulted in a flurry of mergers and acquisitions in the seed and biotech industries. According to the Wallace Centre report, "Relatively few firms control the vast majority of commercial transgenic crop technologies. These firms have strategically developed linkages among the biotechnology, seed, and agri-chemical sectors to capture as much market value as possible. However, these tightly controlled linkages of product sectors raise serious issues of market access, product innovation, and the flow of public benefits from transgenic crops. Unlike Plant Variety Protection—which does not allow for the patenting of individual genes, but only of crop varieties—Intellectual Property Rights prohibit farmers from saving their seed and undertaking their own breeding programme, and prohibit plant breeders from using the material to create a new generation of varieties adapted to specific regions or growing conditions.

Farmers can't save proprietary seed for planting and so must purchase new seed every year. In addition, farmers choosing transgenic varieties must sign a contract with the owner of the variety and pay a substantial per-acre technology fee, or royalty. The anticipated commercial introduction of transgenic wheat represents a dramatic shift in an industry in which farmers still widely save their own seed. As non-transgenic varieties become contaminated with transgenic ones, even those farmers who choose to stick with conventional varieties will lose the right to replant their own seed. This loss has already occurred in Canada's canola industry, with Monsanto winning its court case against farmer Percy Schmeiser for replanting his own canola variety that had become contaminated with Monsanto's Roundup Ready canola.

The adoption of transgenic crop varieties has brought with it an increasing prevalence of contract production. While contract production can lead to increased value and reduced risk for growers, farmers are justified in their concern about their loss of control when they sign a contract with a private company. Issues associated with contract production of transgenic crops must be considered within the broader context of a sustainable agriculture to include ownership, control, and social equity.

GENETIC ENGINEERING

Genetically modified plants are created by the process of genetic engineering, which allows scientists to move genetic material between organisms with the aim of changing their characteristics. All organisms are composed of cells that contain the DNA molecule. Molecules of DNA form units of genetic information, known as genes. Each organism has a genetic blueprint made up of DNA that determines the regulatory functions of its cells and thus the characteristics that make it unique.

Prior to genetic engineering, the exchange of DNA material was possible only between individual organisms of the same species. With the advent of genetic engineering in 1972, scientists have been able to identify specific genes associated with desirable traits in one organism and transfer those genes across species boundaries into another organism. A gene from bacteria, virus, or animal may be transferred into plants to produce genetically modified plants having changed characteristics.

Thus, this method allows mixing of the genetic material among species that cannot otherwise breed naturally. The success of a genetically improved plant depends on the ability to grow single modified cells into whole plants. Some plants like potato and tomato grow easily from single cell or plant tissue. Others such as corn, soy bean, and wheat are more difficult to grow. After years of research, plant specialists have been able to apply their knowledge of genetics to improve various crops such as corn, potato, and cotton. They have to be careful to ensure that the basic characteristics of these new plants are the same as the traditional ones, except for the addition of the improved traits. The world of biotechnology has always moved fast, and now it is moving even faster.

More traits are emerging; more land than ever before is being planted with genetically modified varieties of an ever-expanding number of crops. Research efforts are being made to genetically modify most plants with a high economic value such as cereals, fruits, vegetables, and floriculture and horticulture species.

BIOLOGICAL CELL

It only takes one biological cell to create an organism. In fact, there are countless species of single celled organisms, and indeed multi-cellular organisms like ourselves.

A single cell is able to keep itself functional by owning a series of '*miniature machines*' known as *organelles*. The following list looks at some of these organelles and other characteristics typical of a fully functioning cell.

- *Mitochondrion:* An important cell organelle involved in respiration
- *Cytoplasm:* A fluid surrounding the contents of a cell and forms a vacuole
- *Golgi Apparatus:* The processing area for the creation of a glycoprotein
- *Endoplasmic Reticulum:* An important organelle heavily involved in protein synthesis.
- *Vesicles: Packages* of substances that are to be used in the cell or secreted by it.
- *Nucleus*: The "brain" of a cell containing genetic information that determines every natural process within an organism.
- *Cell Membrane:* Also known as a plasma membrane, this outer layer of a cell assists in the movement of molecules in and out the cell plays both a structural and protective role
- *Lysosomes:* Membranous sacs that contain digestive enzymes

CELL WALL

A structure that characteristically is found in plants and prokaryotes and not animals that plays a structural and protective role.

CELL SPECIALISATION

Cells can become specialised to perform a particular function within an organism, usually as part of a larger tissue consisting of many of the same cells working in tandem:

- *Nerve cells* to operate as part of the nervous system to send messages back and forth via the brain at the centre of the nerve system.
- Skin cells for waterproof protection and protection against pathogens in the open air environment.
- *Xylem* tubes to transport water around plants and to provide structural support for the plant as a whole.

Cells combine their efforts in these tissue types to perform a common cause. The task of the specialised cell will determine in what way it is going to be specialised, because different cells are suited to different purposes, as illustrated in the above list:

- Muscle cells are long and smooth in structure and their elastic nature allows these cells to perform flexible movements, just as they do in our own body's.
- Some *white blood cells* contain powerful digestive enzymes to eliminate pathogens by breaking them down to the molecular level.
- Cells at the back of the eye are sensitive to light stimuli, and thus can *interpret* differences in light intensity which can in turn be interpreted by our nervous system and brain.

Many of these cells contain organelles, though after some cells are specialised, they do not possess particular characteristics as they do not require them to be there. *i.e.* efficiency is the key, no resources are wasted and the resources available are put to their idyllic optimum.

THE CELL MEMBRANE

The *cell membrane*, otherwise known as the plasma membrane is a semi-permeable structure consisting mainly of *phospholipid* (fat) molecules and proteins. They are structured in a *fluid mosaic model*, where a double layer of phospholipid molecules provide a barrier accompanied by proteins.

It is present round the circumference of a cell to acts as a barrier, keeping foreign entities out the cell and its contents (like cytoplasm) firmly inside the cell.

The plasma membrane allows only selected materials to pass in and out of a cell, and is thus known as a selectively permeable membrane.

CELL TRANSPORT

There are three methods in which ions are transported through the cell membrane into the cell,

- *Active Transport*: Active transport is the transport of molecules with the active assistance of a carrier that can transport the material against a natural *concentration gradient*.
- *Passive Transport (Diffusion):* The movement of molecules from areas of high concentration (*i.e.* outside a cell) to areas of low concentration (*i.e.* within a cell) via a carrier. This process does not require energy.
- *Simple Diffusion:* The movement of molecules from areas of high concentration to areas of low concentration in a free state. *Osmosis* of water involves this type of diffusion through a selectively permeable membrane (*i.e.* plasma membrane)

THE BREAKDOWN OF MATERIALS IN A CELL

In cells, sometimes it is required to breakdown more complex molecules into more simple molecules, which can then be 're-built' into what is needed by the body with these new raw materials.

'*Pinocytosis*' where to contents of a structure (such as bacteria) are *drank*, essentially by breaking down molecules into a drinkable form.

'*Phagocytosis*' where contents are 'eaten'.

ABSORPTION AND SECRETION

Absorption is the uptake of materials from a cells' external environment. Secretion is the ejection of material.

BIOLOGICAL ENERGY - ADP AND ATP

ATP stands for Adenosine Tri-Phosphate, and is the energy used by an organism in its daily operations. It consists of an *adenosine* molecule and three inorganic *phosphates*. After a simple reaction breaking down ATP to *ADP*, the energy released from the breaking of a molecular bond is the energy we use to keep ourselves alive.

ATP TO ADP - ENERGY RELEASE

This is done by a simple process, in which one of the phosphate molecules is broken off, therefore reducing the ATP from 3 phosphates to 2, forming ADP (Adenosine Diphosphate after removing one of the phosphates {Pi}). This is commonly wrote as ADP + Pi.

When the bond connecting the phosphate is broken, *energy* is released. While ATP is constantly being used up by the body in its biological processes, the energy supply can be bolstered by new sources of glucose being made available via eating food which is then broken down by the digestive system to smaller particles that can be utilised by the body.

On top of this, ADP is built back up into ATP so that it can be used again in its more energetic state. Although this conversion requires energy, the process produces a net gain in energy, meaning that more energy is available by re-using ADP+Pi back into ATP.

GLUCOSE AND ATP

Many ATP are needed every second by a cell, so ATP is created inside them due to the demand, and the fact that organisms like ourselves are made up of millions of cells.

Glucose, a sugar that is delivered via the bloodstream, is the product of the food you eat, and this is the molecule that is used to create ATP. Sweet foods provide a rich source of readily available glucose while other foods provide the materials needed to create glucose. This glucose is broken down in a series of *enzyme* controlled steps that allow the release of energy to be used by the organism. This process is called respiration.

RESPIRATION AND THE CREATION OF ATP

ATP is created via respiration in both animals and plants. The difference with plants is the fact they attain their food from elsewhere. In essence, materials are harnessed to create ATP for biological processes. The energy can be created via cell respiration.

The process of respiration occurs in 3 steps (when oxygen is present):

1. Glycolysis
2. The Kreb's Cycle
3. The Cytochrome System

CELL RESPIRATION

As mentioned in the previous page on ATP, the process of *respiration* is split into 3 distinct areas that occur at different parts of the cell. Respiration involves the *oxidation* of foodstuff (*i.e.* glucose) in order to create ATP. Respiration can occur with or without oxygen, *aerobic* and *anaerobic* respiration respectively.

GLYCOLYSIS

Glycolysis occurs in the *cytoplasm* of a cell where a 6 carbon glucose molecule (the broken down food that you ate earlier) is broken down by enzymes into a 3 carbon *pyruvic acid*. The execution of this process requires 2 ATP, and produces a net gain of 2 ATP.

The enzymes involved remove hydrogen from the glucose (oxidation) where they take these hydrogen atoms to the cytochrome system, explained soon. In anaerobic respiration, this is where the process ends, glucose is split into 2 molecules of pyruvic acid. When oxygen is present, pyruvic is broken down into other carbon compounds in the Kreb's Cycle. When it is not present, the pyruvic acid is broken down into lactic acid (or carbon dioxide and ethanol).

THE KREB'S CYCLE

When oxygen is present, respiration can harness more ATP from a single unit of glucose. The *pyruvic acid* from the *glycolysis* stage diffuses into a cell organelle called a mitochondrion (pl. *mitochondria*). These mitochondria are sausage shaped structures that host a large surface area for the respiration to occur on.

The pyruvic acid is then subject to more enzymes which break it down into a 2 carbon compound, as seen below. The diagram illustrates the *Kreb's cycle*, consisting of three main actions

- The carbon element is in an infinite cycle where the 2 carbon compound derived from pyruvic acid binds with the 4 carbon compound that is always present in the cycle.
- CO^2 is released, where the oxygen that is present in *aerobic respiration* combines with carbon from the carbon compounds which is released as CO^2. Hence the need for animals to breath out and expel this CO^2.
- Enzymes oxidize the carbon compounds and transport the hydrogen atoms to the cytochrome system.

THE CYTOCHROME SYSTEM

The cytochrome system, also known as the hydrogen carrier system (or the *electron transport system*) are where the reduced hydrogen carriers transport hydrogen atoms from the glycolysis and Kreb's cycle stages. The cytochrome system is found in the many *cristae* of mitochondria, which are tiny stalked particles found on its outer layer. The system contains many 'hydrogen acceptors' which hydrogen can be added to. By following the path of a hydrogen atom, we can see how the cytochrome system works:

- Some *coenzymes* from earlier stages are transferred to the next coenzymes.
- B is then oxidised, therefore the coenzyme releases the hydrogen and energy is made available.
- The released hydrogen atom binds with 2 oxygen atoms (oxygen is available in aerobic respiration) which produces water, a by-product of respiration.

The diagram illustrates this flow of hydrogen within the cytochrome system and how energy is made available by the flow of these atoms. The green circles illustrate where energy is made available via oxidation. Overall their is a gain of 38 ATP from one molecule of glucose in aerobic respiration. The food that we eat provides glucose required in respiration. In plants, energy is also acquired via respiration, but the mechanism of delivering glucose to the respiration process is a little different.

DNA STRUCTURE AND DNA REPLICATION

The energy required by these cells and how energy is created in order for the cell to survive.The structure, type and functions of a cell are all determined by *chromosomes* that are found in the *nucleus* of a cell. These chromosomes are composed of DNA, the acronym for deoxyribonucleic acid.

This DNA determines all the characteristics of an organism, and contains all the genetic material that makes us who we are. This information is passed on from generation to generation in a species so that the information within them can be passed on for the offspring to harness in their lifetime.

STRUCTURE OF DNA AND NUCLEOTIDES

DNA is arranged into a *double helix* structure where spirals of DNA are intertwined with one another continuously bending in on itself but never getting closer or further away. The following diagram illustrates a *nucleotide*, the building blocks of DNA

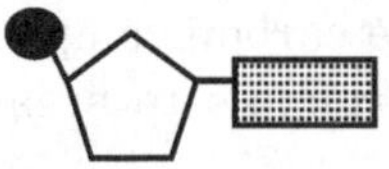

There are four different types of nucleotide possible in a DNA sequence, adenine, cytosine, guanine and thymine (can be replaced with A, C, G and T). There are billions of these nucleotides in our genome, and with all the possible permutations; this is what makes us unique. Nucleotides are situated in adjacent pairs in the double helix nature mentioned.

The following rules apply in regards to what nucleotides pair with one another:

- There are four possible types of nucleotide, adenine, cytosine, guanine and thymine.

- Thymine and adenine can only make up a base pair
- Guanine and cytosine can only make up a base pair
- Therefore, thymine and cytosine would NOT make up a base pair, as is the case with adenine and guanine.

This is illustrated in the below diagram, using correct pairings of nucleotides

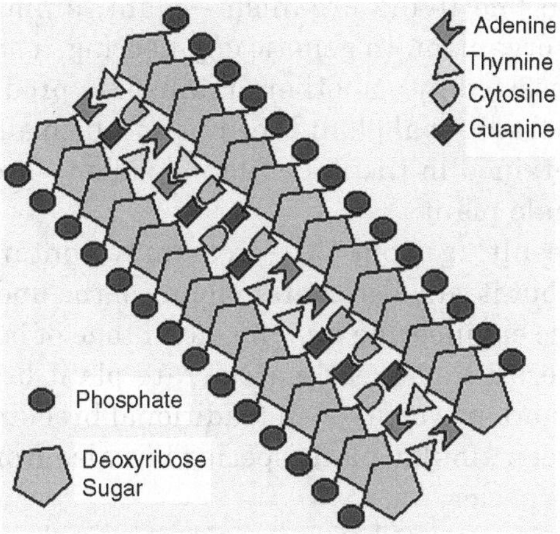

The diagram is two dimensional, remember that DNA is structured in a double helix fashion, as shown to the above right. This continuous sequence, and the sequence they are in determine an organisms' structural, physical and anatomical features.

DNA REPLICATION

Cells do not live forever, and in light of this, they must pass their genetic information on to new cells, and be able to replicate the DNA to be passed on to offspring. It is also required that fragments of DNA (genes) have to be copied to code for particular bodily function. It is essential that the replication of it is EXACT.

In order for replication to occur, the following must be available:

- The actual DNA to act as an exact template
- A pool of relevant and freely available nucleotides
- A supply of the relevant enzymes to stimulate reaction
- ATP to provide energy for these reactions

When replicating, the double helix structure uncoils so that each strand of DNA can be exposed. When they uncoil, the nucleotides are exposed so that the freely available nucleotides can pair up with them. When all nucleotides are paired up with their new partners, they re-coil into the double helix. As there are two strands of DNA involved in replication, the first double helix produces 2 copies of itself via each strand.

It is said that the replicated DNA is semi-conservative, because it possesses 50percent of the original genetic material from its parent. These 2 new copies have the exact DNA that was in the previous one. This template technique allows genetic information to be passed from cell to cell and from parents to offspring.

THE SCIENCE OF MODERN GENETIC ENGINEERING

Genetic engineering is one form of biotechnology and usually refers to copying a gene from one living organism—plant, animal or microbe—and adding it to another organism. In genetic engineering, a small piece of genetic material (DNA) is inserted into another organism to produce a desired effect. This is in contrast to traditional plant breeding in which all the genes desirable and undesirable contained in the male plant—pollen—are combined with all the genes of the female plant.

The progeny resulting from this cross may contain the gene for the desirable character, but it will also contain many of the undesirable genes from both parents. Genetic engineering has the advantage of being able to transfer only the gene of interest and greatly accelerate plant breeding. But genetic engineering also is more powerful than traditional breeding since it can move genes not only between similar plant species but also from distant relatives, including non-plant species.

It is possible to move genes between such seemingly unrelated organisms because all living organisms share the same code for DNA and the synthesis of proteins and other basic life functions. What might seem on the surface to be very different organisms are, in fact, very similar, at least at the molecular level.

All living things are more alike than different, and this is one of the reasons that genes can be moved so successfully between such seemingly different organisms as plants and bacteria. Genes are not unique to the organisms from which they came, so there really aren't "tomato genes" or "bacterial genes." It's the collection of all genes in a tomato or a bacterium that makes it a tomato or bacterium, not a single gene. As we learn more about the genetic makeup of all organisms, we see that most plant species differ by only a small percentage of their genes and those even such seemingly different organisms as tomatoes and bacteria have many common genes.

These findings suggest that in the long-term evolutionary process even tomatoes and bacteria had some common ancestor. From the discovery 50 years ago of the structure of DNA, scientists soon came to realise they could take segments of DNA that carried information for specific traits—genes—and move them into another organism. In 1972, the collaboration of Hubert Boyer and Stanley Cohen resulted in the first isolation and transfer of a gene from one organism to a single-celled bacterium where it would express the gene and manufacture a protein. Their discoveries led to the first direct use of biotechnology—the production of synthetic insulin to treat people with diabetes—and the start of what is often called modern biotechnology. Plants were first transformed through genetic engineering in the late 1970s. Mary-Dell Chilton and colleagues used a common soil-dwelling bacterium, Agrobacterium tumefaciens, that attaches itself to plants and transfers some of its DNA into the plant. Chilton and her colleagues added a gene to this bacterium, which in turn transferred the gene into a plant where it became part of the plant's DNA. This bacterium is still commonly used in genetic engineering along with another technique that uses a high-velocity mechanism to inject DNA into plant cells. The result from either technique is the same—the plant cells take up the gene and begin to express it as their own.

Benefits and Risks of Biotechnology

Plants developed through genetic engineering were first grown on 1.7 million hectares in 1996 in the United States, but by 2002 they were grown on 58.7 million hectares in 16 countries. By far the major use of the present plants is to manage pests—weeds, insects and diseases. Weed management with genetically engineered plants is accomplished because the plants have a modified enzyme (a protein) that allows them to survive an application of a specific herbicide that normally acts on that enzyme. Growers can plant the herbicide-tolerant seeds, allow the plants to emerge along with any weeds in the field and then treat the field with an herbicide.

The result is that the weeds, but not the crops, die. The advantage to growers is that they spend less time on weed management, have enhanced weed control, use safer herbicides, and in many cases use less herbicides. Additionally, this technology allows growers to use soil conservation practices such as reduced or no-tillage, thus helping to retain soil structure and moisture and reduce erosion. Herbicide tolerant crops (soybean, canola, cotton and maize) were grown on 48.6 million hectares in 2002. Insect-resistant crops developed through genetic engineering utilize the common soil bacterium, Bacillus thuringiensis (Bt), which has been commercially used for more than 50 years, as an insecticide spray.

Although safe to humans and the environment, when a susceptible insect ingests Bt, the Bt protein binds to specific molecular receptors in the gut and creates pores causing the insect to starve to death. Insecticidal products containing Bt were first commercialized in France in the late 1930s, but even in 1999 the total sales of Bt products constituted less than 2 per cent of the total value of all insecticides. Bt, which had limited use as a foliar insecticide, became a major insecticide only when genes that produce Bt toxins were engineered into major crops. The Bt crops available at present are maize and cotton. These were grown on a total of 14.5 million hectares in 2002. Virus-resistant crops were created by inserting a non-infective part of a plant virus into a plant, essentially "vaccinating" the plant to protect it from the virus.

This technique is called "pathogen-derived resistance." Squash and papaya have been engineered to resist infection by some common viruses and are approved for sale in the United States. There are fewer than 1 million hectares of these crops. The bioengineered plants available at present provide growers with better tools to manage pest problems. As with any technology, there are risks and benefits to currently available genetically engineered plants, but the present body of information indicates their use has enhanced pest management, substantially reduced the amounts of pesticides used in some crops, enabled growers to use safer pesticides, and contributed to enhanced safety for humans and the environment.

The regulatory process for managing these plants and their effects on the environment and humans has evolved with the technology and the scientific community's knowledge of these tools. Many of the more controversial issues surrounding genetic engineering of plants—such as pesticide resistance, gene flow and intellectual property issues—are not unique to this new technology but pertain to all types of agriculture. Some species of insects have developed resistance to sprays of Bt, indicating the potential for some species to become resistant to Bt plants.

However, despite Bt plants being grown on more than 62 million hectares worldwide from 1996 to 2002, there have been no documented cases of resistance development. The reasons for this lack of resistance appear to involve not only biological factors of the insects and the Bt plant, but also the fact that the regulatory agency (the Environmental Protection Agency) in the United States requires a resistance management plan for growing Bt plants. No other insecticide has such strict regulations. Still, growers, companies and federal regulatory agencies must be vigilant about resistance developing for biotech crops used to manage insects, weeds and viruses as they also must with non-biotech pest management tactics.

It will be important to consider the accrued environmental and health benefits of these biotech crops prior to the development of any resistance and how resistance can be managed if and where it occurs. In addition to pesticide resistance, gene flow from biotech to non-biotech crops may also be a concern. However, the risk of gene flow varies with each crop and each gene. Pollen flow in soybeans is very limited so the risk of a biotech soybean crop crossing with a non-biotech soybean is minimal, but this may be different for another crop. Likewise, if the gene in the biotech crop that provided a pest management trait, such as insect resistance, moved into a non-biotech plant, such as a weed, any selective advantage of the insect-protected weed in the ecosystem should be assessed. These same questions should also be answered with non-biotech crops, but these have not received the same level of attention as biotech crops because of the latter's higher profile.

Horizon of Biotechnology

In the future, the potential uses of plant biotechnology are far more wide-ranging than the pest-management biotech crops of today. Plants are being developed that serve as production "factories" for medically important drugs, sources of alternative energy, tools for cleaning toxic waste sites, and biomaterials including dyes, inks, detergents, adhesives, lubricants, plastics and the like. Consumers may see these products as more directly enhancing their quality of life than the pest-management biotech crops of today. Perhaps an even more dramatic advantage to consumers will be seen when plants are genetically engineered to have enhanced health benefits such as disease-fighting chemicals or increased amounts of essential vitamins and minerals. A healthy and well-informed discussion of the risks and benefits involved in agricultural biotechnology is needed to ensure a proper role for this new technology in our future food and health systems. No one should believe that any technology, including biotechnology, will completely solve the world's agricultural problems. Many people familiar with biotechnology, however, believe it to be an important component of the solution.

GENETIC TREE OF LIFE

In all of biology, there is one thing that always stays the same. That thing, believe it or not, is change itself! The millions of different living things on Earth—plants, bacteria, insects, chimps, people, and everything else—all came to be because of a process called biological evolution, in which organisms change over time.

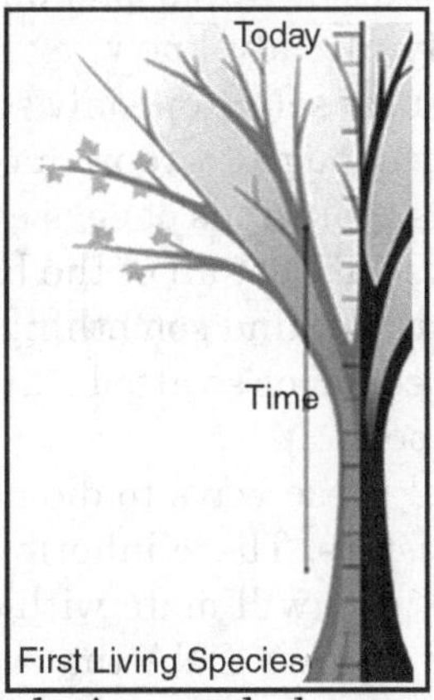

Because of biological evolution, early humans gained the ability to walk on two feet. Because of evolution, air-breathing whales can live in the ocean despite being mammals like us. Because of evolution, some bacteria can live in scalding water, others can survive in solid ice, and still others can live deep in the Earth eating only rocks!

Evolution happens every day, and it affects every species—including us. It changes entire populations, not individuals. And it has a big impact on medical research.

EVERYTHING EVOLVES

To understand evolution, let's go back in time a century and a half to

1854, when the British naturalist Charles Darwin published *The Origin of Species*, a book that proposed an explanation for how evolution works.

The main concept in evolution is that all living things share a common ancestor. The very earliest ancestor of all life forms on Earth lived about 4 billion years ago. From that early organism, millions of types of creatures—some living and some now extinct—have evolved.

Evolution requires diversity. You can tell that living things are diverse just by walking down the street and looking around you. Individual people are very different from one another. Chihuahuas are different from Great Danes, and Siamese cats are different from tabbies.

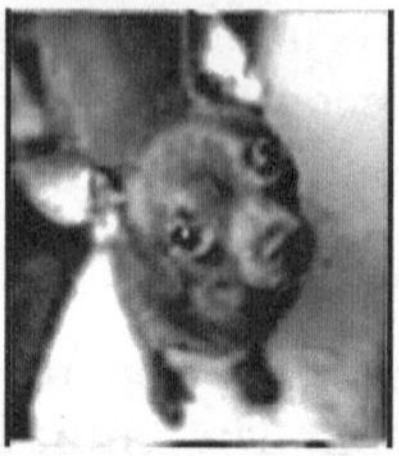

Evolution also depends on inheritance. Many of our unique characteristics are inherited—they are passed from parent to offspring. This is easy to see: Dalmatian puppies look like Dalmatians, not Chihuahuas. Petunias grow differently from pansies. Evolution works *only* on traits that are inherited.

Finally, as you probably already know, evolution favours the "fittest." Through a process called natural selection, only some offspring within a given generation will survive long enough to reproduce. As an example, consider houseflies, each of which lays thousands of eggs every year. Why haven't they taken over the world? Because almost all of the baby houseflies die. The flies that survive are the ones that can find something to eat and drink... the ones that avoid being eaten, stepped on, or swatted... and the ones that don't freeze, drown, or land on a bug zapper.

The flies that survive all these ways to die have what it takes to outlive most of their brothers and sisters. These inherited traits give an organism a survival edge. Those who survive will mate with each other and will pass on to the next generation some of their DNA that encoded these advantageous traits.

Of course, not all aspects of survival are determined by genes. Whether a fly gets swatted depends on genes that affect its reflexes—whether it's fast enough to avoid the swatter—but also on the environment. If there's no human around waving the swatter, the fly is quite likely to survive, regardless of its reflexes.

Evolution often takes a long time to make a difference. But it can also happen very quickly, especially in organisms with short lifespans. For example, as you read earlier, some bacteria have molecular features that let them survive in the presence of antibiotics.When you take an antibiotic medicine, antibiotic-resistant bacteria flourish while antibiotic-sensitive bacteria die.

Because antibiotic resistance is a growing public health threat, it's important to take the whole dosage of antibiotic medicine, not stop when you feel better. And you should take antibiotics only when they're needed, not for colds or other viral infections, which antibiotics can't treat. Viruses must simply run their course.

SELECTIVE STUDY

Scientists doing medical research are very interested in genetic variants that have been selected by evolution. For example, researchers recently discovered a rare genetic variant that protects people from getting AIDS. A genetic variant is a different version of a gene, one that has a slightly different sequence of nucleotides.

Scientists think that the rare variant of a gene called CCR5 originally may have been selected during evolution because it made people resistant to an organism unrelated to HIV.

Montgomery Slatkin of the University of California, Berkeley, has used mathematical modeling techniques to show that natural selection over time could explain the frequency of the CCR5 variant in human populations. The work indicates that the CCR5 gene variant's ability to protect against AIDS may contribute to keeping it in the human gene pool.

So, through evolution, living things change. Sometimes, that's good for us, as when humans understand HIV resistance in hopes of preventing AIDS. But sometimes the changes aren't so great —from a human perspective, anyway—as when bacteria become resistant to antibiotics. Whether the consequences of evolutionary change are good or bad, understanding the process can help us develop new strategies for fighting disease.

CLUES FROM VARIATION

Scientists know quite a bit about how cells reshuffle genetic information to create each person's unique genome. But many things are missing about how this genetic variation contributes to disease, making for a very active area of research.

T C G A T A (A) T G C A T G C A T A —One Person's DNA

T C G A T A (G) T G C A T G C A T A —Another Person's DNA

Different nucleotides (in this example, A or G) can appear in the DNA sequence of the same chromosome from two different individuals, creating a single-nucleotide polymorphism (SNP).

What scientists do know is that most of the human genome is the same in all of us. A little bit of genetic variation—differences that account for much less than 1 per cent of our DNA—gives each of us a unique personality, appearance, and health profile. The parts of the human genome where the

DNA sequences of many individuals vary by a single nucleotide are known as single-nucleotide polymorphisms (abbreviated SNPs and pronounced "snips"). For example, let's say that a certain nucleotide in one of your genes is A. In your uncle, however, the nucleotide in the same place on the same gene might be G. You and your uncle have slightly different versions of that gene. Scientists call the different gene versions alleles.

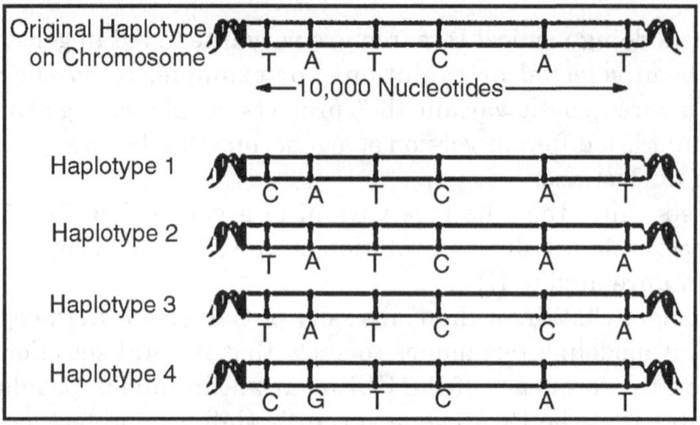

Haplotypes are combinations of gene variants, or SNPs, that are likely to be inherited together within the same chromosomal region. In this example, an original haplotype (top) evolved over time to create three newer haplotypes that each differ by a few nucleotides (red).

If two genes sit right next to each other on a chromosome, the SNPs in those genes tend to be inherited together. This set of neighbouring SNPs is called a haplotype.

Most chromosome regions have only a few, common haplotypes among all humans. As it turns out, these few haplotypes—in different combinations in each person—appear to account for most of the variation from person to person in a population.

Scientists can use haplotype information to compare the genes of people affected by a disease with those of unaffected people. For example, this approach revealed a genetic variation that substantially increases the risk of age-related macular degeneration, the leading cause of severe vision loss in the elderly. Scientists discovered that a single SNP—one nucleotide in the 3 billion-nucleotide human genome—makes some people more likely to get this eye disease. The discovery paves the way for better diagnostic tests and treatments.

What about other diseases? In October 2005, an international scientific team published a catalogue of human haplotypes. Researchers are looking through the catalogue in an effort to identify genes that determine susceptibility to many common diseases, including asthma, diabetes, cancer, and heart disease.

But not all SNPs are in genes. Scientists studying genetic variation have also found SNPs in DNA that doesn't encode proteins. Nonetheless, some of these SNPs appear to affect gene activity. Some researchers suspect that the "cryptic" (hidden) variation associated with SNPs in non-coding DNA plays an important role in determining the physical characteristics and behaviours of an organism.

Loren Rieseberg of Indiana University in Bloomington is one scientist who would love to take the mystery out of cryptic variation. He wants to know how this non-coding genetic variation can help organisms adapt to new environments. He's also curious about whether it can create problems for some individuals.

You might be surprised to learn that Rieseberg's principal research subject is the sun- flower. Although many plants produce only one generation a year, plants like sunflowers can be very useful tools for researchers asking fundamental questions about genetics. Because their genetic material is more malleable than that of many animals, plants are excellent models for studying how evolution works. Wild sunflowers appealed to Rieseberg because there are several species that live in different habitats. Two ancient species of wild sunflowers grow in moderate climates and are broadly distributed throughout the central and western United States.

Fig. Plants like these Sunflowers make Great Models for Studying how Evolution works.

Three recently evolved sunflower species live in more specialized environments: One of the new species grows on sand dunes, another grows in dry desert soil, and the third species grows in a salt marsh.

To see how quickly new plant species could evolve, Rieseberg forced the two ancient sunflowers to interbreed with each other, something plants but not other organisms can do. Among the hybrid progeny were sunflowers that were just like the three recently evolved species! What that means is that Rieseberg had stimulated evolution in his lab, similar to what actually happened in nature some 60,000 to 200,000 years ago, when the newer species first arose.

That Rieseberg could do this is pretty amazing, but the really interesting part is how it happened. Scientists generally assume that, for a new species with very different characteristics to evolve, a lot of new mutations have to occur.

But when Rieseberg looked at the genomes of his hybrid sunflowers, he was surprised to find that they were just cut-and-pasted versions of the ancient sunflower species' genomes: large chunks had been moved rather than many new SNPs created. Rieseberg reasons that plants stash away unused genetic material, giving them a ready supply of ingredients they can use to adapt quickly to a new environment. It may be that human genomes can recycle unused genetic material to confront new challenges, as well.

LIVING LABOURATORIES

The Living Labouratories, which is also available as a poster. To order a free copy of the poster,

THE GENOME ZOO

Scientists often use an image of a tree to depict how all organisms, living and extinct, are related to a common ancestor. In this "tree of life," each branch represents a species, and the forks between branches show when the species represented by those branches became different from one another. For example, researchers estimate that the common ancestor of humans and chimpanzees lived about 6 million years ago. While it is obvious just by looking that people have a lot in common with our closest living relatives, chimpanzees, what about more distant species? If you look at an evolutionary tree, you'll see that humans are related to mice, worms, and even bacteria. The ancestral species that gave rise to both humans and bacteria was alive a lot longer ago than the ancestor of humans and chimpanzees, yet we still share hundreds of genes with bacteria. Scientists use the term comparative genomics to describe what they're doing when they compare the genomes of different species to see how similar (or how different!) the species' DNA sequences are. Sequences that the species have in common are the molecular footprints of an ancestor of those species.

Why are "old" DNA sequences still in our genomes? It turns out that nature is quite economical, so DNA sequences that are responsible for something as complicated and important as controlling gene activity may stay intact for millions of years. Comparative genomic studies also have medical implications. What would you do if you wanted to develop new methods of preventing, diagnosing, or treating a human disease that animals don't get?

If people have a gene that influences their risk for a disease, and mice have the gene too, you could study some aspect of the disease in mice, even though they don't ever have the symptoms of the disease. You could even study the disease in yeast, if it has the gene, as well.

STARTING ALL OVER AGAIN

Stem cells—what embryos are made up of just days after an egg is fertilized by a sperm—have the amazing ability to grow up into any kind of cell: skin, heart, muscle, nerve, and everything else. How do they do it?

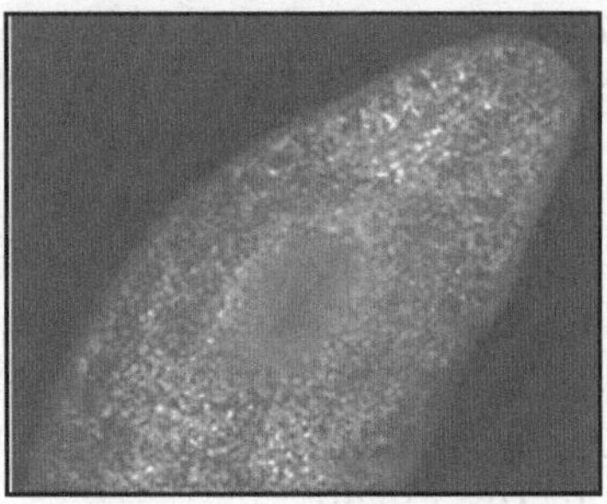

Fig. Phillip Newmark

Intrigued by the potential of these masterful cells, researchers want to know just what it is that gives stem cells their ability to change into anything, upon the body's request, but stay in the "I can do anything" state until asked.

Some basic researchers are trying to figure out how stem cells work by using a unique model system: tiny, freshwater worms called planarians. These worms are like stem cells in the sense that they can regenerate. You can cut up planarians into hundreds of pieces, and each piece will regenerate into an intact worm that looks the same as all the others.

Planarians' resemblance to stem cells isn't just coincidence. Scientists have discovered that planarians can perform the amazing act of regeneration due to the presence of, yes, specialized stem cells in their bodies.

Developmental biologist Alejandro Sánchez Alvarado of the University of Utah School of Medicine in Salt Lake City used the gene-silencing technique RNAi to search for planarian genes that were essential for regeneration. He found 240 genes that, when silenced, caused a physical defect in the worm's growth and regenerative ability. Interestingly, 16 per cent of these looked very much like genes that had been linked to human disease!

Sánchez Alvarado and his team hope to figure out how regeneration genes allow the specialized stem cells within the worm to travel to a wounded site and "turn into" any of the 30 or so cell types needed to recreate a mature worm. Although humans are only distantly related to planarians, we have many of the same genes, so these findings have the potential to reveal strategies for regenerating injured body parts in people, too.

GENES MEET ENVIRONMENT

If toxins from the environment get into our bodies, they don't always make us sick. That's because liver enzymes come to our rescue to make the chemicals less harmful. The genes that encode those enzymes are under constant evolutionary pressure to adapt quickly to new toxins.

For example, certain liver enzymes called cytochrome P450 proteins metabolize, or break down, hormones that our bodies make as well as many of the foreign substances that we encounter. These include harmful molecules like cancer-causing agents as well as beneficial ones, like medicines. In fact, just two genes within the cytochrome P450 family, abbreviated 3A4 and 3A5, encode proteins that process more than half of all of the medicines that are sold today.

Since the chemicals to which people are exposed vary so widely, a scientist might predict that there would be different variants of cytochrome P450 genes in different human populations. Using comparative genomics, researchers such as Anna Di Rienzo of the University of Chicago have shown that this is indeed the case. Di Rienzo has found many sequence differences within these genes in people living throughout the world.

It turns out that one variant of the gene that encodes the cytochrome P450 3A5 protein makes this enzyme very efficient at breaking down cortisol, a hormone that raises salt levels in the kidneys and helps the body retain water. Di Rienzo compared the DNA sequences of the 3A5 gene in DNA samples taken from more than 1,000 people representing over 50 populations worldwide. She was amazed to find a striking link between the existence of the gene variant and the geographic locale of the people who have it.

Di Rienzo discovered that African populations living very close to the equator were more likely than other populations to have the salt-saving version of the 3A5 gene. She suggests that this is because this gene variant provides a health advantage for people living in a very hot climate, since retaining salt helps ward off dehydration caused by intense heat.

However, there seems to be a cost associated with that benefit—the 3A5 gene variant raises the risk for some types of high blood pressure. That means that in environments in which retaining salt is not beneficial, evolution selects against this gene variant.

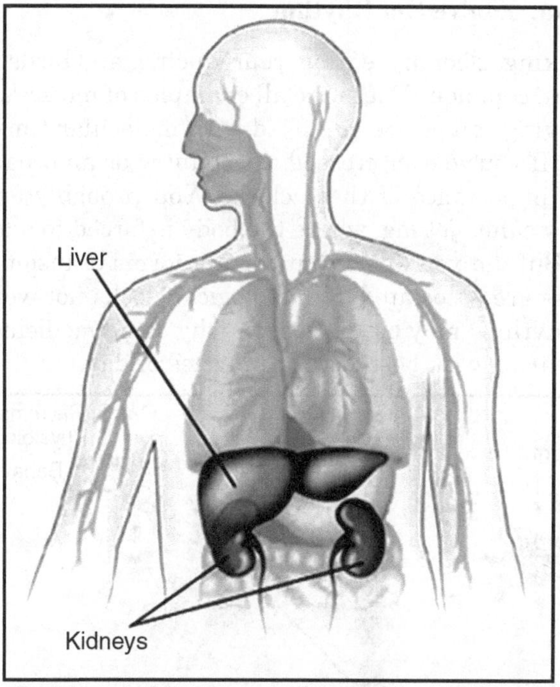

Fig. The Liver and Kidneys are Susceptible to Damage from Toxins since these body Organs Process Chemicals.

Another scientist who studies interactions between genes and the environment is Serrine Lau of the University of Arizona in Tucson. She studies a class of harmful molecules called polyphenols, present in cigarette smoke and car exhaust, that cause kidney cancer in rats, and perhaps, in people.

Lau discovered that rats and humans who are more sensitive to some of the breakdown products of polyphenols have an unusual DNA sequence-a genetic signature-that increases their risk of developing cancer. She suspects that the gene that is affected encodes a tumor suppressor: a protein that prevents cancer from developing. In people and rats with the genetic signature, she reasons, the tumor suppressor doesn't work right, so tumors grow.

Taking this logic one step further, it may be that certain people's genetic make-up makes them unusually susceptible to DNA damage caused by exposure to carcinogens. If doctors could identify those at risk, Lau says, such people could be forewarned to avoid contact with specific chemicals to protect their health.

However, think about this scenario: Who should make those decisions? For example, would it be ethical for an employer to refuse to hire somebody because the person has a genetic signature that makes him or her more likely to get cancer if exposed to a chemical used in the workplace? Tough question.

Genetics and You: You've Got Rhythm

What do waking, sleeping, eating, reproducing, and birds flying south for the winter have in common? These are all examples of nature's amazing sense of rhythm. All living things are equipped with molecular timepieces that set the pulse of life. If you've ever crossed the country or an ocean by plane, you know about the importance of these clocks. You probably experienced that traveller's misery called jet lag, where the body is forced to adapt quickly to a new time zone. But did you know that certain forms of insomnia and manic-depressive illness are associated with biological clocks not working properly? And biological rhythms may be the reason why some medicines and surgical treatments appear to work best at certain times of day.

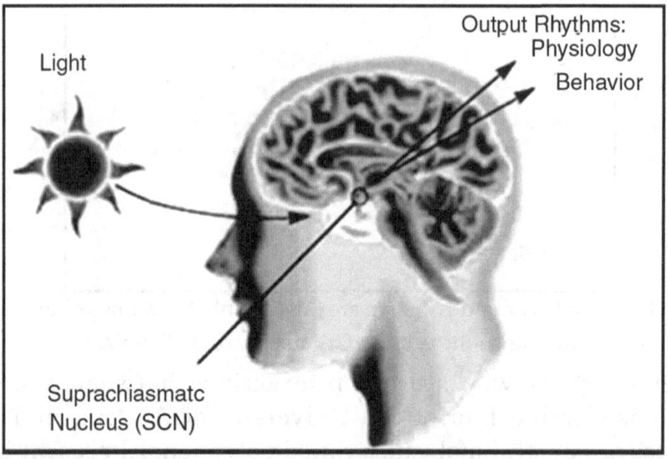

The human body keeps time with a master clock called the suprachiasmatic nucleus or SCN. Situated inside the brain, it's a tiny sliver of tissue about the size of a grain of rice, located behind the eyes. It sits quite close to the optic nerve, which controls vision, and this means that the SCN"clock" can keep track of day and night. Given enough time, your SCN can reset itself after you fly in an airplane from one time zone to another.

The SCN helps control sleep by coordinating the actions of billions of miniature"clocks" throughout the body. These aren't actually clocks, but rather are ensembles of genes inside clusters of cells that switch on and off in a regular, 24-hour cycle-our physiological day. Scientists call this 24-hour oscillation a circadian rhythm. ("Circadian" comes from the Latin words meaning"approximately a day.") Researchers have discovered that all living things-plants, animals, and bacteria-have circadian rhythms. Many researchers working with insect and other model systems have identified genes that are critical for keeping biological time.

Understanding circadian rhythms will help scientists better understand sleep disorders. If we have the opportunity, most of us sleep 7 or 8 hours at night, and if we don't get enough rest we may have a hard time getting things done the next day. Some people, however, routinely get by with only 3 to 4 hours of sleep. Researchers have noted that this trait seems to run in families, suggesting a genetic link. As it turns out, fruit flies need even more sleep than people. Neuroscientist Chiara Cirelli of the University of Wisconsin-Madison did a genetic search for fruit fly mutants that don't sleep much. She discovered that flies with a variant of a gene called *shaker* sleep only 3 to 4 hours per night.

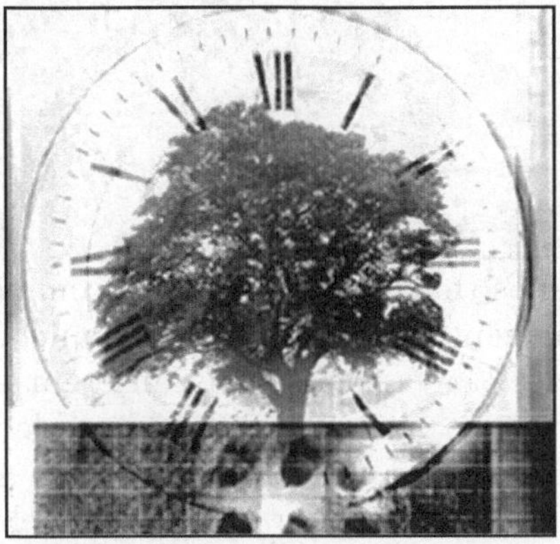

Although the *shaker* flies don't appear sleep-deprived, Cirelli found that they have a different problem: They don't live as long as flies without the mutation. She is now studying this new connection between sleep and lifespan. Her work may also pave the way for improved sleep aids and effective remedies for jet lag.

ANIMALS HELPING PEOPLE

Using technology that grew out of the Human Genome Project, scientists have read the sequences of the genomes of hundreds of organisms: dogs, mice, rats, chickens, honeybees, fruit flies, sea urchins, puffer fish, sea squirts, roundworms, several fungi, and many disease-causing bacteria. Next in line are dozens of additional species including marmosets, skates, tree shrews, sloths, and zebra finches!

What effect will all this gene sequence information have on medical research? We've already talked about the fact that people share many of their genes with other species. This means that when scientists read the sequence of another species' genome, they're likely to discover that the organism has many of the genes that, in humans, cause disease or raise disease risk when mutated.

Take fruit flies as one example. According to biologist Ethan Bier of the University of California, San Diego, 30 per cent of the currently identified human disease genes most likely have functional counterparts in none other than *Drosophila melanogaster*, a fruit fly species widely used in genetic research. Currently, Bier and other scientists are using experimental flies to investigate a wide range of genes involved in conditions such as blindness, deafness, mental retardation, heart disease, and the way in which bacterial toxins cause illness. By reading the DNA sequences of many other species, researchers hope to find model systems that are even better than fruit flies for studying some aspects of human disease.

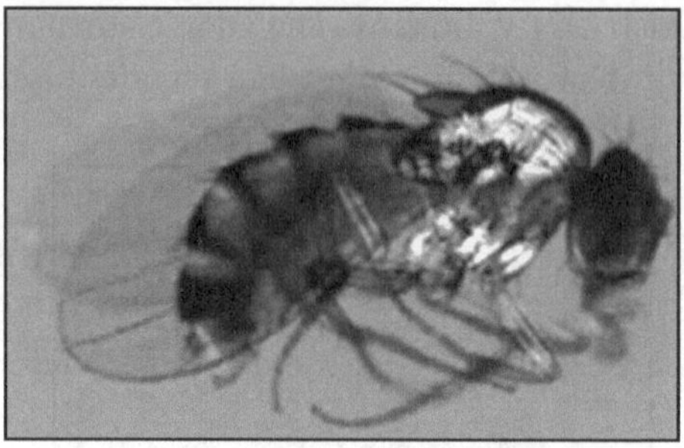

Sometimes, the genes that we don't have in common with other species are as important as the genes we share. For example, consider the fact that humans and chimpanzees have remarkably different abilities and physical features. But the chimpanzee genome is 99 per cent identical to our own. And did you know that chimpanzees don't get malaria or AIDS? So about 1 per cent of our genome determines whether we look and behave like a person or a chimp, and whether we are susceptible to malaria or AIDS. Rather, the gigantic task of comparing the nucleotides that make up the genomes of two or more species is the perfect job for a computer, a natural multitasker. If you consider that the human genome contains 3 billion nucleotides, you can easily see why this is work well suited to a machine (with a human operator, of course).

Researchers called computational biologists help analyse genomic data. These scientists develop software programmes that enable computers to perform genome comparisons. Among other things, the programmes can figure out where in the DNA sequences a gene starts and stops: its "boundaries."

Other researchers who work in the field of bioinformatics mine genomic information hidden in the masses of data. They are looking for scientific treasure in the form of new biological knowledge. These experiments can zero in on previously hidden patterns and reveal links between different fields of research. Bioinformaticists and computational biologists are in high demand because they play a very important role in 21st-century medical science. These scientists must be fluent in both computer science and biology.

Bibliography

A.G. Ponniah and George John: *Fish Chromosome Atlas*, National Bureau of Fish Genetic Resources, 1998.

A.K. Sharma and Ramavtar Sharma: *Crop Improvement And Mutation Breeding*, Scientific Publication, Jaipur, 2014.

Ashok K. Singh: *Records of the Zoological Survey of India : A Catalogue of Experimental Studies on Grasshopper Chromosomes in India 1928 to 2000*, ZSI, 2002.

B.B. Buchanan: *Biochemistry and Molecular Biology of Plants*, Wilhelm Gruissem and Russell L. Jones, I.K. International, Delhi, 2007.

Biological Control of Plant Pathogens, Weeds and Phytoparasitic Nematodes : *M. Reddi Kumar, M. John Sudheer, B. Chenchu Reddy and B. Srinivasulu*, B.S. Publication, Delhi, 2008.

D. Thangadurai, T. Pullaiah and Pedro A. Balatti: *Genetic Resources and Biotechnology, Vol. 2*, Regency Publication, Delhi, 2005.

Dharmendra Singh and S. Manivannan: *Genetic Resources of Horticultural Crops*, International Book Distributing Co, Delhi, 2009.

Hari Har Ram and Rakesh Yadava: *Genetic Resources and Seed Enterprises : Management and Policies (2 Vols-Set)*, New India Publishing Agency, Delhi, 2007.

Hema Gautam: *Genetic Modification of Plants*, Rajat Publication, Delhi, 2006.

K.D. Srivastava: *Biology and Management of Wheat Pathogens*, Studium Press, Delhi, 2010.

L.V. Gangawane and V.C. Khilare: *Molecular Biology of Plant Pathogens*, Daya Publication, Delhi, 2010.

Laurence M. Winters: *Animal Breeding*, Greenworld Publication, Delhi, 2002.

M M Ranga: *Transgenic Animals*, Agrobios Publication, Jaipur, 2006.

P. Parvatha Reddy: *Biofumigation and Solarization for Management of Soil Borne Plant Pathogens*, Scientific Publication, Delhi, 2011.

P. Vidhyasekaran: *Bacterial Disease Resistance in Plants : Molecular Biology and Biotechnological Applications*, International Books, Delhi, 2005.

P.V.G.K. Reddy: *Genetic Resources of Indian Major Carps*, Daya Publication, Delhi, 2005.

Pragya Khanna: *Cell and Molecular Biology*, I K International, Delhi, 2008.

Prakash S Lohar: *Cell and Molecular Biology*, MJP Publication, Delhi, 2007.

R C Sharma and J N Sharma: *Plant Diseases, Pathogens and Their Vectors : An Annotated List*, Westville Publication, New Delhi, 2008.

Ranjeet K. Singh: *Genetic Resources and Biotechnology*, Surendra Publication, Delhi, 2010.

S.D. Sharma: *Genetic Resources of Rice in India : Past and Present*, Today and Tomorrows Publication, Delhi, 2010.

Sheela Srivastava: *Genetic Modified Crops and Genetic Engineering*, Manglam Publication, Delhi, 2011.

T.K. Chandra Bhanu and Vasudev Bhatnagar: *Genetic Science and Technology*, Campus Books, Delhi, 2009.

V.K. Prabhakar: *Genetic Species and Ecosystem Diversity*, Anmol Publication, Delhi, 2001.

Vladimir I. Vysotskii and Alla A. Kornilova: *Nuclear Transmutation of Stable and Radioactive Isotopes in Biological Systems*, Pentagon Press, London, 2010.

William H. Elliott and Daphne C. Elliott: *Biochemistry and Molecular Biology*, Oxford University Press, Delhi, 2005.

Y.S. Ahlawat: *Diagnosis of Plant Viruses and Allied Pathogens*, Studium Press, Delhi, 2010.

Yashodhara Dalmia and Salima Hashmi: *Memory Metaphor Mutations : Contemporary Art of India and Pakistan*, Oxford University Press, Delhi, 2007.

Index